Construction Management

The application of quality tools and techniques in construction projects has a great influence on the cost-effectiveness results of construction projects and achieving successful project performance. Quality management tools and techniques help in project planning, execution, monitoring, and control of the project and evolve a project management system that makes project deliverables.

Construction Management: Quality Tools and Techniques provides the usage and application of various quality tools and techniques in different phases of construction project management focusing on three quality management processes involving the varying aspects of quality. This volume illustrates how important quality is for construction projects and offers an overview of construction projects along with a brief introduction to the different types of projects, the different phases of the life cycle, and the principles. Project Delivery Systems (PDS) along with their organizational relationship, participants, advantages, and disadvantages of each system are included. The book goes on to discuss the qualification of construction managers and the type of management systems and their roles during the different design stages from the beginning through to the handover of the project is also included.

This volume provides significant information and guidelines to construction and project management professionals (owners, designers, consultants, construction managers, project managers, supervisors, contractors, builders, developers, and many others from the construction-related industry) involved in construction projects (mainly civil construction projects, commercial-A/E projects) and construction-related industries.

Systems Innovation Book Series

Series Editor: Adedeji Badiru

Systems Innovation refers to all aspects of developing and deploying new technology, methodology, techniques, and best practices in advancing industrial production and economic development. This entails such topics as product design and development, entrepreneurship, global trade, environmental consciousness, operations and logistics, introduction and management of technology, collaborative system design, and product commercialization. Industrial innovation suggests breaking away from the traditional approaches to industrial production. It encourages the marriage of systems science, management principles, and technology implementation. Particular focus will be the impact of modern technology on industrial development and industrialization approaches, particularly for developing economics. The series will also cover how emerging technologies and entrepreneurship are essential for economic development and society advancement.

Project Management for the Oil and Gas Industry
A Global System Approach, Second Edition, 2-Volume Set
Adedeji B. Badiru

Project Management for the Oil and Gas Industry
Systems Engineering and Technology, Volume 1
Adedeji B. Badiru

Project Management for the Oil and Gas Industry
Agile Systems and Environmental Concerns, Volume 2
Adedeji B. Badiru

Handbook of Construction Management
Scope, Schedule, and Cost Control, 2-Volume Set
Edited by Abdul Razzak Rumane

Construction Management
Quality Tools and Techniques
Edited by Abdul Razzak Rumane

Construction Management
Project Management, Process Principles, and Concepts
Edited by Abdul Razzak Rumane

Industrial Engineering Strategy for Constructive Technologies
A Systems-Based Approach for the Global Economy
Adedeji B. Badiru

Construction Management

Quality Tools and Techniques

Second Edition

Edited by
Abdul Razzak Rumane

CRC Press
Taylor & Francis Group
Boca Raton London New York

CRC Press is an imprint of the
Taylor & Francis Group, an **informa** business

Designed cover image: Shutterstock and Abdul Razzak Rumane

Second edition published 2026
by CRC Press
2385 NW Executive Center Drive, Suite 320, Boca Raton FL 33431

and by CRC Press
4 Park Square, Milton Park, Abingdon, Oxon, OX14 4RN

CRC Press is an imprint of Taylor & Francis Group, LLC

First edition published by CRC Press 2016

ISBN: 978-1-041-06696-5 (hbk)
ISBN: 978-1-041-06697-2 (pbk)
ISBN: 978-1-003-63668-7 (ebk)

DOI: 10.1201/9781003636687

Typeset in Times
by Apex CoVantage, LLC

To
My Parents
For their prayers and love
My prayers are always for my father and my mother who encouraged and inspired me all the times.
I wish they would have been here to see this book and give me blessings.

To
My Wife
I miss my wife, who stood with me all the time during writing of my earlier books.

Contents

Preface

It's almost seven years passed since my book entitled *Handbook of Construction Projects: Scope, Schedule, and Control* was published. The response was encouraging, and many people appreciated the book as it was totally devoted to providing significant information and guidelines to construction and project management professionals (owners, designers, consultants, construction managers, project managers, supervisors, and many others from construction-related industry), involved in construction projects (mainly civil construction projects, commercial A-E projects), and construction related industries. It covers the importance of construction management principles, procedures, concepts, methods, and tools and their applications to various activities/components/subsystems of different phases of the life cycle of a construction project to improve the construction process in order to conveniently manage the project and make the project most qualitative, competitive, and economical. I am happy and feel honored that I could share my knowledge which helped many of the readers in their day-to-day professional activities related to quality in construction projects.

Since then, ISO 31000 has been revised and ISO 45000 is already published. There were certain quality management-related topics and some other construction management principles were missing in the first edition, which I feel are important to have comprehensive information about construction management. Further, some of the figures and tables need to be updated to make a comprehensive understanding of the subject.

Due to the voluminous nature of construction management processes and with the addition of certain missing topics, and for proper understanding of the required information the book is divided into two volumes mainly consisting of topics related to quality tools and techniques and processes, systems for execution of construction project.

Volume 1 is titled *Construction Management: Quality Tools and Techniques* and is, divided into nine chapters and numbered sections.

Volume 2 is titled *Construction Management: Project Manage Process Principles and Concepts* and is divided into five chapters and numbered sections.

Quality in construction projects is achieved through the application of various quality control principles, procedures, concepts, methods, tools, and techniques and their applications to various activities/components/subsystems at different phases of the life cycle of a construction project to improve construction process to conveniently manage the project and make them more qualitative, competitive, and economical.

Quality tools and techniques are very important to develop a comprehensive project quality management system. The application of tools and techniques in construction projects has a great influence on cost-effectiveness results of construction projects and achieving successful project performance. Quality management tools and techniques help in project planning, execution, monitoring, and control of project and evolve a project management system that makes project deliverables:

- Meet defined scope
- Completed as per schedule
- Completed within budget

The book has been developed to provide significant information about the usage and applications of various quality tools and techniques in different phases of construction projects focusing on three quality management processes namely: plan quality, perform quality assurance, and perform quality control.

For the sake of proper understanding, the book is divided into nine chapters and each chapter is divided into a number of sections covering quality-related topics that have importance or relevance for understanding quality and project management concepts for construction projects.

Chapter 1 is about Overview of Construction Projects. It will cover a brief introduction about types of construction projects, different phases of the construction project life cycle, and principles of quality in construction projects.

Chapter 2 is about Project Delivery Systems (PDS). It discusses different types of project delivery systems and their organizational relationship among various project participants, advantages, and disadvantages of each of these systems. It also discusses different types of contracting systems based on pricing methods.

Chapter 3 is about Construction Management Delivery System. It discusses the qualification of construction managers and the types of construction management systems (agency CM and CM at risk). It also discusses the roles of the construction manager during the pre-design stage, design stage, during construction, and post-construction (testing, commissioning, and handover).

Chapter 4 is about Construction Management Tools. It gives a brief description of various types of quality tools in practice, mainly by the construction industry, such as Classic Tools of Quality, Management and Planning Tools, Process Analysis Tools, Process Improvement Tools, Innovation and Creative Tools, Lean Tools, Cost of Quality, Quality Function Deployment, Six Sigma, Application of Six Sigma Tools in construction projects, and TRIZ. The usage of each of the tools under these categories is supported with a table, figure, and chart to make an easy understanding about their applications in construction projects. The chapter also includes quality information systems. It also discusses briefly quality assessment/measures and customer satisfaction.

Chapter 5 is about Building Information Modeling (BIM) in Design and Construction. This chapter provides brief information about the use of BIM as a collaborative tool in construction projects, BIM to manage complex projects, and BIM execution plans.

Chapter 6 is about Construction Contract Documents. It covers brief information about different types of contract documents used to prepare construction documents.

Chapter 7 is about ISO Certification in Construction Industry. It covers brief information about the importance of standards and standardization bodies. It covers a case study detailing ISO implementation methodology detailing the implementation of QMS, documentation, and certification. The chapter includes brief information about ISO 14001 and ISO 27001. The chapter also includes a correlation matrix between ISO 9001:2008 and ISO 9001:2015 and the contents of the quality management system manual for designers and contractors.

Chapter 8 is about Quality Auditing standards and auditing processes for construction projects.

Chapter 9 is about Value Methodology. It discusses value engineering-related basic topics such as the objectives of value engineering, timings of value engineering, and the benefits of value engineering are also included in this topic.

The book, I am certain, will meet the requirements of construction professionals, students, and academics and satisfy their needs.

Foreword

A lot has changed since Dr. Rumane's first edition of this book. This second edition has been both necessary after seven years of existence and required to update the essential parts of the process. The author has introduced additional concepts and tools not included in the first edition. This second edition is written by Dr. Abdul Rumane again as a compendium of tools and a detailed methodology for completing a construction project in an efficient and cost-effective manner. The logic and sequence of the chapters lend themselves to assist the modern construction manager, which includes students, professors, and practitioners to understand the requirements of today's complex and demanding construction environment.

This second edition of *Handbook of Construction Management: Scope, Schedule, and Cost Control*—now split into two volumes (*Quality Tools and Techniques* and *Construction Management: Project Management, Process Principles, and Concepts*)—is also an extension of Dr. Rumane's previous books. In this second edition, the construction community is provided with additional and updated management advice and concrete examples to establish and maintain quality during all aspects of the project life cycle.

All practitioners in the construction business can use this new edition to improve both their own internal as well as external construction management processes and practices. Included in this edition are numerous updated figures and tables that continue to support his understanding of construction management.

Dr. Rumane's own continued personal experiences and well-developed understanding allow him to improve the first edition into a much fuller and complete handbook for the construction manager and staff. He continues to explore the additional quality tools, management techniques, and practices used by leading construction companies in the industry of the construction business. The information presented in this second edition will give the reader a continued competitive edge when it comes to construction management processes, maintaining quality and effectively operating throughout the life cycle of a construction project.

It is always a pleasure to discuss these subjects, and I value Dr. Rumane's professional and systemic approach to construction. As a retired quality engineer, I continue to stay involved in the quality business and hope this second edition will allow you to enjoy this next journey to learning and improving your own construction management knowledge. I know this book will provide you with additional tools to make your continuing journey a rewarding one.

Raymond R. Crawford,
Retired
Past Chair, Design and Construction Division, ASQ
Previous certificates included CMQ O/E, CQA, LEED AP
September 2024

Acknowledgment

Share the knowledge with others is the motto of this book.

Many of my colleagues and friends extended help while preparing the book by arranging reference material, many thanks to all of them for their support.

I thank the publishers and authors, whose writings are included in this book, for extending their support by allowing me to reprint their material.

I thank reviewers from various professional organizations, for their valuable input to improve my writing. I thank members of ASQ Audit Division, ASQ Design and Construction Division, The Institution of Engineers (India), IEI, Kuwait Chapter, Kuwait Society of Engineers, and ASQ Kuwait Section, for their support to bring out this book.

I thank Raymond R. Crawford for his nicely worded thought provocative Foreword and support and best wishes.

I thank Cindy Renee Carelli, Executive Editor of CRC Press, Senior Editorial Assistant, and other CRC staff for their support and contribution to make this construction-related book a reality.

My sincere thanks to the following contributor authors and co-authors for their contribution to the book:

- Lincoln Forbes
- Cliff Moser
- Jitendra Patel
- Zofia Rybkowski
- Mustafa Shraim
- Edward Taylor

I thank Dr. Adedeji Badiru of the Airforce Institute of Technology, Mr. Collin Sutt, Former Chair, ASQ Design and Construction Division, Mr. Cliff Moser, Former Chair, ASQ Design and Construction Division, Mr. Hasan Salih, Chair, ASQ Design and Construction Division for their best wishes all the time.

I thank Engr. Adel Kharafi, Former Chairman Kuwait Society of Engineers, and former President of the World Federation of Engineering Organizations (WFEO) for his good wishes all the time, I thank Engr. Ahmad Alkandari, Director, Kuwait Municipality for his support and good wishes. I thank Engr. Ahmad Almershed, former Undersecretary, MSNA, Kuwait for his good wishes all the time. I thank Dr. Ayed Alamri, President, Saudi Quality Council, KSA for his support and good wishes all the time. I thank Dr. Fadel Safer, former Minister of Public Works, Kuwait for his support and good wishes. I thank Engr. Faisal D. Alatal, Chairman, Kuwait Society of Engineers and President of the Federation of Arab Engineers for his support and good wishes. I thank Prof. Mohammed Aichouni, University of Hail, KSA for his support and good wishes all the time. I thank Dr. Mohammad Ben Salamah, former Chair, ASQ Kuwait Section for his support and best wishes. I thank Dr. N. N. Murthy of Jagruti Kiran Consultants for his support and good wishes. I thank

Dr. Othman Alshamrani, Imam Abdulrehman Bin Faisal University, KSA for his support and good wishes all the time. I thank Maj. Gen. R. K. Sanan, VSM, (Retd), former Secretary Director General, The Institution of Engineers (India) for his support and best wishes all the time. I thank Ms. Rima Al Awadhi former Chair, ASQ Kuwait Section and Team Leader at Kuwait Oil Company for her support and good wishes. I thank Engr. Wael Aljasem of Kuwait Project Management Society for his support and good wishes. I thank Engr. Yaseen Farraj, former Director, Ministry of Public Works Kuwait for his support and good wishes. I thank Engr. Waleed Al Mutawa, General Manager KSE for his support and good wishes.

I thank Mr. Bashir Ibrahim Parkar of Dar SSH International, Engr. Ashraf Hajwane, Engr. Annama Issac, Engr. Asif Kadiwala of KNPC, and Engr. Ateeq Mirza of KOC for their valuable input and support.

I extend my thanks to Dr. Ted Coleman, Professor and Department Chair, California State University, San Bernardino, and former Chancellor KW University, for his everlasting support.

My special thanks to H.E. Sheikh Rakan Nayef Jaber Al Sabah for his support and good wishes all the time.

I thank my well-wishers whose inspiration made me complete this book.

Most of the data discussed in this book is from the author's practical and professional experience and is accurate to the best of the author's knowledge and ability. However, if any discrepancies are observed in the presentation, I would appreciate communicating them to me.

The contribution of my son Ataullah, my daughter Farzeen, and daughter-in-law Masum is worth mentioning here. They encouraged me and helped me in my preparatory work to achieve the final product. I thank my brothers, sisters, and all the family members for their support, encouragement, and good wishes all the time.

Abdul Razzak Rumane

About the Editor

Abdul Razzak Rumane PhD is Chartered Quality Professional-Fellow of the Chartered Quality Institute (UK) and a KSE Certified consultant engineer in the field of electrical engineering and project management. He obtained a bachelor of engineering (electrical) degree from Marathwada University (now Dr. Babasaheb Ambedkar Marathwada University) India in 1972 and received his Doctor of Philosophy (PhD) from Kennedy Western University, USA (now Warren National University) in 2005. His dissertation topic was "Quality Engineering Applications in Construction Projects."

Dr. Rumane has been honored by The International University of Ministry and Education, Missouri, USA, which awarded the honorary degree of Doctor of Philosophy (D.Phil.) for expertise in quality management in 2021, and the Yorker International University, USA, awarded an honorary doctorate of engineering in 2007. The Albert Schweitzer International Foundation honored him with a gold medal for "Outstanding contribution in the field of Quality in Construction Projects." World Quality Congress awarded him the "Global Award for Excellence in Quality Management and Leadership."

Dr. Rumane is an accomplished engineer. He is associated with a number of professional organizations. He is a fellow of The Institution of Engineers (India) and has an honorary fellowship of the Chartered Management Association (H.K.), a senior member of ASQ, a member of PMI, and a member of KSE, among others.

Dr. Rumane has attended many international conferences and has made technical presentations at various conferences. Dr. Rumane is the author of nine useful books entitled *Quality Management in Construction Projects*, first edition (2010) *Quality Tools for Managing Construction Projects* (2013), *Quality Management in Construction Projects*, second edition (2017), *Quality Management in Oil and Gas Projects* (2021), *Risk Management Applications Used to Sustain Quality in Projects: A Practical Guide* (2022), *Quality Management: How to Achieve Sustainability in Projects* (2023), *Total Quality Management Applications and Concepts for Construction Projects* (2024), and editor of the book titled *Handbook of Construction Management: Scope, Schedule, and Cost Control* (2016). All of these books are published by CRC Press (Taylor & Francis Group), USA. A book entitled *Quality Auditing in Construction Projects: A Handbook* (2019) is published by Routledge, UK (Taylor & Francis Group). His book *Quality Management in Construction Projects* is translated into Korean language.

Presently he is International Liaison Committee Chair ASQ (Design and Construction Division). He also served in this position in the year 2023–24. He earlier served as Treasurer of ASQ Kuwait Section for the year 2022, Secretary ASQ GC for the year 2019 and Secretary ASQ LMC, Kuwait for the years 2017 and 2018. He served as honorary chairman of The Institution of Engineers (India), Kuwait chapter for the years 2016–2017, 2013–2014, and 2005–2007.

Dr. Rumane is a trainer in the field of quality in construction projects, construction management, and project management.

Dr. Rumane's professional career exceeds 50 years including ten years in manufacturing industries and over 40 years in construction projects. Presently he is associated with SIJJEEL Co., Kuwait as advisor and director, construction management.

Contributors

Cliff Moser
Self Employed
Berkeley, California, USA

Cliff Moser, AIA, is a licensed architect with over 45 years of extensive experience across a diverse range of construction types. With more than 25 years dedicated to California OSHPD healthcare projects, Cliff has excelled in roles both as an architect and an owner's representative. Despite initially leaving architecture school, he achieved licensure through hands-on experience with major international architectural firms.

Further solidifying his expertise, Cliff earned a master of science in quality assurance. His knowledge led him to author the Architectural Practice Quality Management chapters in the 14th and 15th editions of the *AIA Handbook of Professional Practice*. As the operations director for an international BIM consulting firm, he trained AECO organizations in leveraging BIM throughout the entire design and construction lifecycle.

Currently a doctoral candidate in complex systems science, Cliff has authored numerous papers and presentations, including *Architecture 3.0: The Disruptive Design Practice Handbook*. Today, he applies his expertise as a forensic architect and standard of care professional. Connect with Cliff at CliffMoser.com.

Mustafa Shraim
Associate Professor, College of Engineering
and Technology, Ohio University
Athens, Ohio, USA

Mustafa Shraim is an associate professor in the Department of Engineering Technology and Management at Ohio University in Athens, Ohio. He received his PhD in industrial engineering from West Virginia University. He teaches courses in quality management systems, statistical quality control, and leadership. His research interests are in quality engineering, quality management systems (QMS), and Dr. W Edwards Deming's theories of management—also known as *System of Profound knowledge*™

Mustafa has over 25 years of industrial experience as a quality engineer, corporate quality manager, and a consultant in quality management systems, statistical methods, and quality improvement. He facilitated Lean projects and mentored project leaders in Value Stream Mapping, Kaizen (A3) methodology, and 5S systems. Mustafa is a Certified Quality Engineer (CQE) and Six Sigma Black Belt (SSBB) by ASQ, and a certified QMS Principal Auditor by IRCA. He is also an ASQ Fellow.

Abbreviations

AAMA	American Architectural Manufacturers Association
ACI	American Concrete Institute
AISC	American Institute of Steel Construction
AMCA	American Composite Manufacturers Association
ANSI	American National Standards Institute
API	American Petroleum Institute
ARI	American Refrigeration Institute
ASCE	American Society of Civil Engineers
ASHRAE	American Society of Heating, Refrigeration and Air-conditioning Engineers
ASTM	American Society of Testing Materials
ASQ	American Society for Quality
BMS	Building Management System
BREEAM	Building Research Establishment Environmental Assessment
BSI	British Standard Institute
CDM	Construction (Design and Management)
CEN	European Committee for Standardization
CIBSE	Chartered Institution of Building Services Engineers
CIE	International Commission on Illumination
CII	Construction Industry Institute
CMAA	Construction Management Association of America
CSC	Construction Specifications, Canada
CSI	Construction Specification Institute
CTI	Cooling Tower Industry
DIN	Deutsches Institute fur Normung
EIA	Electronic Industry Association
EJCDC	Engineering Joint Contract Document Committee
EN	European Norms
FIDIC	Federation International des Ingeneurs-Counceils
HQE	Higher Quality Environmental
ICE	Institute of Civil Engineers (UK)
IEC	International Electrotechnical Commission
IEEE	Institute of Electrical and Electronics Engineers
IP	Ingress Protection
ISO	International Organization for Standardization
LEED	Leadership in Energy and Environmental Design
NEC	National Electric Code
NEC	New Engineering Contract
NEMA	National Electrical Manufacturers Association (USA)
NFPA	National Fire Protection Association
NWWDA	National Wood, Window and Door Association
PMI	Project Management Institute

PMBOK	Project Management Book of Knowledge
QS	Quantity Surveyor
RFID	Radio Frequency Identification
SDI	Steel Door Institue
TIA	Telecommunication Industry Association
UL	Underwriters Laboratories

Synonyms

Owner	Client, Employer
Consultant	Architect/Engineer (A/E), Designer, Design Professionals, Consulting Engineers, Supervision Professional
Engineer	Resident Project Representative
Engineer's Representative	Resident Engineer
Project Manager	Construction Manager (Agency CM)
Contractor	Constructor, Builder
Quantity Surveyor	Cost Estimator, Contract Attorney, Cost Engineer, Cost and Works Superintendent
Main Contractor	General Contractor

Figures

Tables

1 Overview of Construction Projects

Abdul Razzak Rumane

1.1 CONSTRUCTION PROJECTS

Construction is the translation of the owner's goals and objectives into a facility built by the contractor/builder as stipulated in the contract documents, plans, and specifications on schedule and within the budget. Construction projects are a custom rather than a routine, repetitive business and differ from manufacturing. Construction projects work against a defined scope, schedule, and budget to achieve the specified result.

Construction has a history of several thousand years. The first shelters were built from stone or mud, and the materials were collected from the forests to provide protection against cold, wind, rain, and snow. These buildings were primarily for residential purposes, although some may have commercial functions.

During the New Stone Age, people introduced dried bricks, wall construction, metalworking, and irrigation. Gradually people developed the skills to construct villages and cities and considerable skills in building were acquired. This can be seen from the great civilizations in different parts some 4000–5000 years ago. During the early period of Greek settlements, which was about 2000 BC, the buildings were made of mud using timber frames. Later temples and theatres were built from marble. Some 1500–2000 years ago, Rome became the leading center of world culture which extended to construction.

Marcus Pollo was a first-century BC military and civil engineer, who published books in Rome. This was the world's first major publication on architecture and construction, and it dealt with building materials, the styles and designs of building types, the construction process, building physics, astronomy, and building machines.

During the Medieval Age (476–1492), improvements in agriculture and artisanal productivity, exploration, and consequent broadening of commerce took place and in the late Middle Ages building construction became a major construction industry. Craftsmen were given training and education in order to develop skills and to raise their status. At this time, Guilds were responsible to manage the quality.

The fifteenth century brought a "renaissance" or renewal in architecture, building, and science. Significant changes occurred during the seventeenth century and thereafter due to the increasing transformation of construction and urban habitat.

The scientific revolution of the seventeenth and eighteenth centuries gave birth to the great Industrial Revolution of the eighteenth century. After some delay, construction followed these developments in the nineteenth century.

DOI: 10.1201/9781003636687-1

The first half of the twentieth century witnessed the construction industry becoming an important sector throughout the world, employing many workers. During this period skyscrapers, long-span dams, shells, and bridges were developed to satisfy new requirements and marked the continuing progress of construction techniques. The provision of services such as heating, air conditioning, electrical lighting, mains water, and elevators to buildings became common. The twentieth century has seen the transformation of the construction and building industry into a major economic sector. During the second half of the twentieth century, the construction industry began to industrialize, introducing mechanization, prefabrication, and system building. The design of building services systems changed considerably in the last 20 years of the twentieth century. It became the responsibility of designers to follow health, safety, and environmental regulations while designing any building.

Residential and commercial, traditional A/E types of construction projects account for an estimated 25% of the annual construction volume. Building construction is a labor-intensive endeavor. Every construction project has some elements that are unique. No two construction or R&D projects are alike. Though it is clear that projects are usually more routine than research and development projects, some degree of customization is a characteristic of the projects.

There are several types of projects. TABLE 1.1 illustrates types of construction projects.

A project is a plan or program performed by people with assigned resources to achieve an objective within a finite duration. A project is a temporary endeavor undertaken to create a unique product or service. "Temporary" means that every project has a definite beginning and a definite end. Projects have specified objectives to be completed within certain specifications and funding limits. Projects are often critical components of the performing organization's business strategy. Examples of projects include:

- Developing a new product or service.
- Effecting a change in structure, staffing, or style of an organization.
- Designing a new transportation vehicle/Aircraft.
- Developing or acquiring a new or modified information system.
- Running a campaign for political office.
- Implementation of a new business procedure or process.
- Constructing a building or facilities.

The duration of a project is finite, projects are not ongoing efforts and the project ceases when its declared objectives have been attained. Some of the characteristics of projects, for example are:

1. Performed by people.
2. Constrained by limited resources.
3. Planned, executed, and controlled.

Construction projects comprised of a cross section of many different participants. These participants both influence and depend on each other in addition to "other

TABLE 1.1
Types of Construction Projects

1	**Process Type Projects**			
1.1	Liquid chemical plants			
1.2	Liquid/solid plants			
1.3	Solid process plants			
1.4	Petrochemical plants			
1.5	Petroleum refineries			
2	**Non-Process Type Projects**			
2.1	Power plants			
2.2	Manufacturing plants			
2.3	Support facilities			
2.4	Miscellaneous (R&D) projects			
2.5	Civil construction projects	**Categories of Civil construction projects and Commercial A/E projects**	Residential construction	Family homes, multi-unit townhouses, gardens, apartments, condominiums, high-rise apartments, villas.
2.6	Commercial A/E projects		Building construction (institutional and commercial)	Schools, universities, hospitals, commercial office complexes, shopping malls, banks, theatres, stadiums, government buildings, warehouses, recreation centers, amusement parks, holiday resorts, neighborhood centers.
			Industrial construction	Petroleum refineries, petroleum plants, power plants, heavy manufacturing plants, steel mills, chemical processing plants.
			Heavy engineering	Dams, tunnels, bridges, highways, railways, airports, urban rapid transit systems, ports, harbors, power lines, communication networks.
			Environmental	Water treatment and clean water distribution, sanitary and sewage system, waste management.

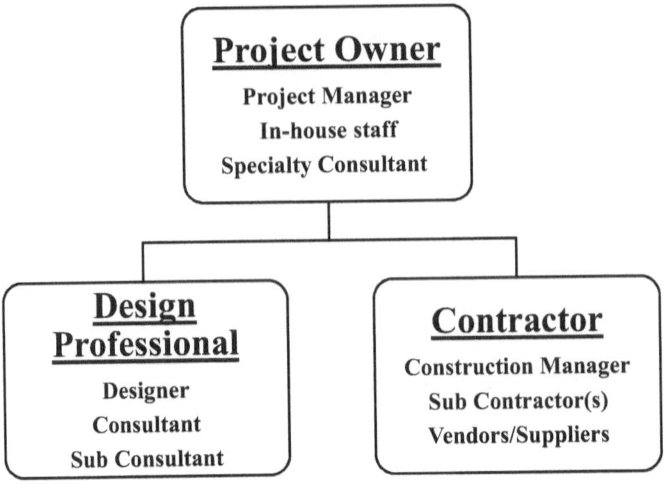

FIGURE 1.1 Concept of traditional project organization.

players" involved in the construction process. FIGURE 1.1 illustrates the concept of traditional construction project organization (traditional contracting system).

Traditional construction projects have involvement of three main groups. These are:

1. Owner—A person or an organization that initiates and sanctions a project. He/she requests the needs of the facility and is responsible to arrange the financial resources for creation of the facility.
2. Designer (A/E)—This consists of architects or engineers or consultants. They are the Owner's appointed entity accountable to convert the owner's conception and need into the specific facility with detailed directions through drawings and specifications within the economic objectives. They are responsible for the design of the project and in certain cases supervision of the construction process.
3. Contractor—A construction firm engaged by the owner to complete the specific facility by providing the necessary staff, workforce, materials, equipment, tools, and other accessories to the satisfaction of the owner/end user in compliance with the contract documents. The contractor is responsible for implementing the project activities and to achieve the owner's objectives.

Construction projects are executed based on a predetermined set of goals and objectives. Under traditional construction projects, the owner heads the team, designated as a project manager. The project manager is a person/member of the owner's staff or independently hired person/firm who has overall or principal responsibility for the management of the project as a whole.

In certain cases, owners engage a professional firm, called the construction manager, who is trained in the management of construction processes, to assist in developing bid documents, overseeing, and coordinating the project for the owner. The basic construction management concept is that the owner assigns a contract to a firm that is knowledgeable and capable of coordinating all the aspects of the project to meet the intended use of the project by the owner. In construction management type of construction projects, the consultant (Architect/Engineer) prepares complete design drawings and contract documents then the project is put for competitive bid and the contract is awarded to the competitive bidder (contractor). The owner hires a third party (construction manager) to oversee and coordinate the construction.

Construction projects are mainly capital investment projects. They are customized and non-repetitive in nature. Construction projects have become more complex and technical, and the relationships and the contractual grouping of those who are involved are also more complex and contractually varied. The products used in construction projects are expensive, complex, immovable, and long-lived. Generally, a construction project is composed of building materials (civil), electro-mechanical items, finishing items, and equipment. These are normally produced by other construction-related industries/manufacturers. These industries produce products as per their own quality management practices complying with certain quality standards or against specific requirements for a particular project. The owner of the construction project or his/her representative has no direct control over these companies unless he/his representative/appointed contractor commits to buying their product for use in their facility. These organizations may have their own quality management program. In manufacturing or service industries the quality management of all in-house manufactured products is performed by the manufacturer's own team or under the control of the same organization having jurisdiction over their manufacturing plants at different locations. Quality management of vendor-supplied items/products is carried out as stipulated in the purchasing contract as per the quality control specifications of the buyer.

1.2 CONSTRUCTION PROJECT LIFE CYCLE

Systems are as pervasive as the universe in which we live. The world in which we live may be divided into the natural world and the human-made world. Systems appeared first in natural forms and subsequently with the appearance of human beings, human-made system systems came into existence. Natural systems are those that came into being by natural process. Human-made systems are those in which human beings have intervened through components, attributes, or relationships.

The systems approach is a technique, which represents a broad-based systematic approach to problems that may be interdisciplinary. It is particularly useful when problems are affected by many factors, and it entails the creation of a problem model that corresponds as closely as possible in some sense to reality. The systems approach stresses the need for the engineer to look for all the relevant factors, influences, or components of the environment that surround the problem. The systems approach corresponds to a comprehensive attack on a problem and to an interest in, and commitment to, formulating a problem in the widest and fullest manner that can be professionally handled.

Systems engineering and analysis when coupled with new emerging technologies, reveal unexpected opportunities for bringing new improved systems and products into being that will be more competitive in the world economy. Product competitiveness is desired by both commercial and public sector producers worldwide. It is the product or consumer good that must meet customer expectations.

These technologies and processes can be applied to construction projects. Systems engineering approach to construction projects helps understand the entire process of project management in order to understand and manage its activities at different levels of various phases to achieve economic and competitive results. The cost-effectiveness of the resulting technical activities can be enhanced by giving more alteration to what they are to do, before addressing what they are composed of. To ensure economic competitiveness regarding the product, engineering must become more closely associated with economics and economic facilities. This is best accomplished through the life cycle approach to engineering. Every system is made up of components and components can be broken down into similar components. If two hierarchical levels arc involved in a given system, the lower is conveniently called a subsystem. The designation of system, sub-system, and components are relatives because the system at one level in the hierarchy is the component at another level.

Most construction projects are custom-oriented, having a specific need and a customized design. It is always the owner's desire that his/her project should be unique and better. Further, it is the owner's goal and objective that the facility is completed on time. The expected time schedule is important from both the financial and acquisition of the facility by the owner/end user.

The system life cycle is fundamental to the application of systems engineering. Systems engineering approach to construction projects helps understand the entire process of project management and to manage and control its activities at different levels of various phases to ensure timely completion of the project with economical use of resources to make the construction project most qualitative, competitive, and economical.

Systems engineering starts from the complexity of the large-scale problem as a whole and moves towards structural analysis and partitioning process until the questions of interest are answered. This process of decomposition is called a Work Breakdown Structure (WBS). The WBS is a hierarchical representation of system levels. Being a family tree, the WBS consists of a number of levels, starting with the complete system at level 1 at the top and progressing downward through as many levels as necessary, to obtain elements that can be conveniently managed.

Benefits of systems engineering applications are:

- Reduction in cost of system design and development, production/construction, system operation and support, system retirement, and material disposal.
- Reduction in system acquisition time.

- More visibility and reduction in the risks associated with the design decision-making process.

Though, it is difficult to generalize the project life cycle to the system life cycle. However, considering that there are innumerable processes that make up the construction process, the technologies and processes, as applied to systems engineering, can also be applied to construction projects. The number of phases shall depend on the complexity of the project. Duration of each phase may vary from project to project. Generally, the major construction projects (non-process type of projects) have seven most common phases. These are as follows:

1. Conceptual Design.
2. Schematic Design.
3. Design Development.
4. Construction Documents
5. Bidding and Tendering
6. Construction.
7. Testing, Commissioning, and Handover.

Each phase can further be subdivided into the Work Breakdown Structure (WBS) principle to reach a level of complexity where each element/activity can be treated as a single unit that can be conveniently managed. Work breakdown structure represents a systematic and logical breakdown of the project phase into its components (activities). It is constructed by dividing the project into major elements and with each of these being divided into sub-elements. This is done till a breakdown is done in terms of manageable units of work for which responsibility can be defined. WBS involves envisioning the project as a hierarchy of goals, objectives, activities, subactivities, and work packages. The hierarchical decomposition of activities continues until the entire project is displayed as a network of separately identified and non-overlapping activities, Each activity will be single-purposed, of a specific time duration, and manageable, its time and cost estimates easily derived, deliverables clearly understood, and responsibility for its completion clearly assigned. The work breakdown structure helps in:

- Effective planning by dividing the work into manageable elements that can be planned, budgeted, and controlled.
- Assignment of responsibility for work elements to project personnel and outside agencies.
- Development of control and information system.

Work breakdown structure facilitates the planning, budgeting, scheduling, and control activities for the project manager and their team. By application of the work breakdown structure phenomenon, the construction phases are further divided into various activities. Division of these phases will improve the control and planning of the construction project at every stage before a new phase starts. The components/

activities of construction project lifecycle phases divided on the WBS principle are listed:

1. Conceptual Design
    - Identification of need.
    - Feasibility.
    - Project goals and objectives.
    - Identification of alternatives/options.
    - Selection of preferred activity.
    - Project delivery/contracting system.
    - Selection of project team (designer A/E).
    - TOR/project charter.
    - Development of concept design.
    - Identification of stakeholders.
    - Establish concept design scope of work.
    - Prepare concept design.
    - Prepare time schedule.
    - Estimate project cost.
    - Quality management.
    - Estimate resources.
    - Identify/manage risk.
    - HSE issues/requirements.
    - Review concept design.
    - Finalize concept design.

2. Schematic/Preliminary Design
    - Identification of stakeholders/project team.
    - Identification of preliminary design requirements.
    - Establish preliminary design scope.
    - Develop preliminary management plan.
    - Develop preliminary design.
    - Develop contract terms and conditions.
    - Regulatory/authorities' approval.
    - Estimate preliminary schedule.
    - Estimate project cost.
    - Quality management.
    - Estimate resources.
    - Identification/management of risk.
    - HSE issues/requirements.
    - Perform value engineering study.
    - Review preliminary design.
    - Finalize schematic design.

3. Detailed Design
    - Identification of stakeholders/project team.
    - Identification of detailed design requirements.

- Develop detailed design of the works.
- Regulatory/authorities' approval.
- Prepare contract documents and specifications.
- Prepare schedule.
- Estimate cost.
- Quality management.
- Estimate resources.
- Identification/management of risk.
- HSE issues/requirements.
- Finalize detailed design.

4. Construction Documents
   - Identification of stakeholders/project team members.
   - Identification of construction documents requirements.
   - Establish construction documents scope.
   - Develop construction documents.
   - Develop tender documents.
   - Estimate project schedule.
   - Estimate cost/budget.
   - Quality management.
   - Estimate resources.
   - Identification/management of risk.
   - HSE issues/requirements.
   - Review construction documents (designer).
   - Review construction documents (owner).
   - Finalize construction documents.
   - Release for bidding and tendering.

5. Bidding and Tendering
   - Identification of stakeholders/project team members.
   - Organize tender documents.
   - Identification of tendering procedure.
   - Identification of bidders.
   - Manage tendering process.
   - Submit/receive tender documents.
   - Manage bidding and tendering quality.
   - Identification/management of risk.
   - Review bid documents.
   - Select contractor.
   - Award contract.

6. Construction
   - Identification of stakeholders/project teams.
   - Mobilization.
   - Development of project site facilities.
   - Identification of project execution/instruction requirements.

- Identification of sustainability requirements.
- Develop project execution scope.
- Planning and scheduling.
- Construction/execution of works.
- Monitoring and control.
- Inspection of executed/installed works.
- Validate execute works.

7. Testing, Commissioning, and Handover
    - Identify stakeholders.
    - Identify testing, commissioning, and handover requirements.
    - Identify sustainability requirements.
    - Develop testing and commissioning scope.
    - Develop inspection and testing plan.
    - Execute commissioning works/systems.
    - Manage testing, commissioning quality.
    - Develop documents.
    - Train owner's/end user's personnel.
    - Handover of facility to owner/end user.
    - Issue substantial completion certificate.
    - Lesson learned.
    - Settle payments.
    - Settle claim.
    - Close contract.

TABLE 1.2 illustrates sub-divided activities, components of construction project life cycle phases.

These activities may not be strictly sequential; however, the breakdown allows the implementation of project management functions more effectively at different stages.

1.3 PRINCIPLES OF QUALITY IN CONSTRUCTION PROJECTS

Quality has different meanings to different people. The definition of quality relating to manufacturing, processes, and service industries is as follows:

- Meeting the customer's needs.
- Customer satisfaction.
- Fitness for use.
- Conforming to requirements.
- Degree of excellence at an acceptable price.

The International Organization for Standardization (ISO) defines quality as "the totality of characteristics of an entity that bears on its ability to satisfy stated or implied needs."

TABLE 1.2
Construction Project Lifecycle Phases

Construction Project Life Cycle (Design-Bid-Build) Phases

Conceptual Design	Schematic Design/Preliminary Design	Design Development/Detail Design	Construction Documents	Bidding and Tendering	Construction	Testing, Commissioning, and Handover
• Identification of Need	• Identification of Stakeholders/Project Team Members	• Identification of Stakeholders/Project Team Members	• Identification of Stakeholders/Project Team Members	• Identification of Stakeholders/Project Team Members	• Identify Stakeholders/Project Teams	• Identify Stakeholders/Project Team members
• Feasibility	• Identification of Preliminary Design Requirements	• Identification of Detailed Design Requirements	• Identification of Construction Documents Requirements	• Organize Tender Documents	• Mobilization	• Identify Testing, Commissioning, and Handover Requirements
• Project Goals and Objectives		• Establish Detailed Design Scope	• Establish Construction Documents Scope	• Identify Tendering Process	• Development of Project Site Facilities	• Identify Sustainability Requirements
• Identification of Alternatives/Options	• Develop Management Plan	• Develop Project Management Plan	• Develop Tender Documents	• Identify Bidders	• Identification of Project Execution/Installation Requirements	• Develop Testing, Commissioning Scope
• Select Preferred Alternative	• Develop Preliminary Design	• Develop Detailed Design	• Develop Tender Documents	• Establish Bid Review Process	• Identification Sustainability Requirement	• Develop Inspection and Testing Plan

(Continued)

TABLE 1.2 (Continued)

Construction Project Lifecycle Phases

• Establish Project Delivery and Contracting System	• Prepare Contract Terms and Conditions	• Develop Contract Documents	• Develop Project Schedule	• Manage Tendering Process	• Establish Project Execution/ Installation Scope	• Execute Commissioning of Works/ Systems
• Selection of Project Team (Designer)	• Regulatory/ Authority Approval	• Regulatory Approval	• Estimate Project Cost/Budget	• Submit/Receive Bids	• Project Planning and Scheduling	• Manage Testing, Commissioning Quality
• Terms of Reference (TOR)/Project Charter	• Estimate Preliminary Schedule	• Prepare Schedule	• Manage Construction Documents Quality	• Manage Bidding and Tendering Quality	• Develop Management Plans	• Develop Documents
• Development of Concept Design	• Estimate Project Cost	• Estimate Cost	• Estimate Resources	• Identification/ Management of Tendering Risk	• Construction/ Execution of Works	• Train Owner's/ End User's Personnel
• Identify Concept Design Stakeholders	• Quality Management	• Manage Design Quality	• Identification/ Management of Risk	• Review Bid Documents	• Monitoring and Control	• Handover Project
• Establish Concept Design Scope of work	• Estimate Resources	• Estimate Resources	• HSE Issues/ Requirements	• Select Contractor	• Inspection of Executed Works/Systems	• Issue Substantial Completion Certificate
• Prepare Concept Design	• Identification/ Management of Risk	• Identification/ Management of Risk	• Review Construction Documents (Designer)	• Award Contract	• Validate Executed Works	• Lesson Learned
• Develop Time Schedule	• HSE Requirements	• HSE Issues/ Requirements	• Review Construction Documents (Owner)			• Settle Payments

(Continued)

TABLE 1.2 (Continued)
Construction Project Lifecycle Phases

• Estimate Project Cost	• Perform Value Engineering Study	• Review Detailed Design	• Finalize Construction Documents		• Settle Claims
• Quality Management	• Review Preliminary Design	• Finalize Detailed Design	• Release for Bidding and Tendering		• Close Contract
• Estimate Resources	• Finalize Preliminary Design				
• Management of Risk/ HSE Issues/ Requirements • Review Concept Design • Finalize Concept Design					

However, the definition of quality for construction projects is different to that of manufacturing or services industries as the product is not repetitive, but a unique piece of work with specific requirements.

Quality in a construction project is not only the quality of product and equipment used in the construction of the facility, it is the total management approach to complete the facility. The quality of construction depends mainly upon the control of construction, which is the primary responsibility of the contractor.

Quality in manufacturing passes through a series of processes. Material and labor are input through a series of processes out of which a product is obtained. The output is monitored by inspection and testing at various stages of production. Any nonconforming product identified is either repaired, reworked, or scrapped, and proper steps are taken to eliminate problem causes. Statistical process control methods are used to reduce variability and to increase the efficiency of processes. In construction projects, the scenario is not the same. If anything goes wrong, the nonconforming work is very difficult to rectify, and remedial actions are sometimes not possible.

Quality management in construction projects is different to that of manufacturing. Quality in construction projects is not only the quality of products and equipment used in the construction, it is the total management approach to complete the facility as per the scope of work to customer /owner satisfaction within the budget and to be completed within specified schedule to meet owner's defined purpose. The nature of the contracts between the parties plays a dominant part in the quality system required from the project and the responsibility for achieving them must therefore be specified in the project documents. The documents include plans, specifications, schedules, bill of quantities, and so on. Quality control in construction typically involves ensuring compliance with minimum standards of material and workmanship in order to ensure the performance of the facility according to the design. These minimum standards are contained in the specification documents. For the purpose of ensuring compliance, random samples and statistical methods are commonly used as the basis for accepting or rejecting work completed and batches of materials. Rejection of a batch is based on non-conformance or violation of the relevant design specifications.

Based on the aforementioned, quality of construction projects can be defined as: Construction project quality is the fulfillment of the owner's needs as per the defined scope of work within a budget and specified schedule to satisfy the owner's/ user's requirements. The phenomenon of these three components can be called the "Construction Project Trilogy" and is illustrated in FIGURE 1.2.

Thus, quality of construction projects can be evolved as follows:

1. Properly defined scope of work.
2. Owner, project manager, design team leader, consultant, and constructor's manager are responsible to implement the quality.
3. Continuous improvement can be achieved at different levels as follows:
    a. Owner specifies the latest needs.
    b. Designer specification to include the latest quality materials, products, and equipment.
    c. Constructor uses the latest construction equipment to build the facility.

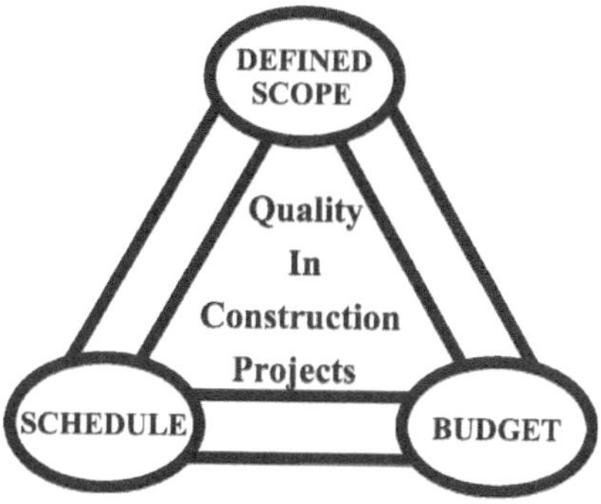

FIGURE 1.2 Construction project trilogy.

4. Establishment of performance measures.
   a. Owner—1) To review and ensure that designer has prepared the contract documents which satisfy his/her needs.
      2) To check the progress of work to ensure compliance with the Contract documents.
   b. Consultant—1) As a consultant designer to include the owner's requirements explicitly and clearly defined in the contact documents.
      2) As a supervision, consultant supervise contractor's work as per contract documents and the specified standards.
   c. Contractor—To construct the facility as specified and use the materials, products, and equipment that satisfy the specified requirements.

5. Team Approach—Every member of the project team should know the principles of Total Quality Management (TQM) knowing that TQM is a collaborative effort, and everybody should participate in all the functional areas to improve the quality of project work. They should know that it is a collective effort by all the participants to achieve project quality.
6. Training and Education—Consultants and contractors should have customized training plans for their management, engineers, supervisors, office staff, technicians, and laborers.
7. Establish Leadership—Organizational leadership should be established to achieve the specified quality. Encourage and help the staff and laborers to understand the quality to be achieved for the project.

These definitions when applied to construction projects relate to the contract specifications or owner/end user requirements to be constructed in such a way that construction of the facility is suitable for the owner's use, or it meets the owner's requirements. Quality in construction is achieved through a complex interaction of many participants in the facilities development process.

The quality plan for construction projects is part of the overall project documentation consisting of:

1. Well-defined specifications for all the materials, products, components and equipment to be used to construct the facility.
2. Detailed construction drawings.
3. Detailed work procedure.
4. Details of the quality standards and codes to be complied.
5. Cost of the project.
6. Manpower and other resources to be used for the project.
7. Project completion schedule.

Participation involvement of all three parties at different levels of construction phases is required to develop a quality system and application of quality tools and techniques. With the application of various quality principles, tools, and methods by all the participants at different stages of a construction project, rework can be reduced resulting in savings in the project cost and making the project qualitative and economical. This will ensure completion of construction and make the project most qualitative, competitive, and economical. TABLE 1.3 illustrates the quality principles of construction projects.

TABLE 1.3
Quality Principles of Construction Projects

Principle	Construction Projects' Quality Principle
Principle 1 (Customer Focus)	1.1 Designer (consultant) is responsible for providing owner's requirements explicitly and clearly defining the standards of the end products and their compliance in the contract documents.
	1.2 Engineering design should include the process, process equipment, engineering systems requirements clearly and without any ambiguity for ease of operation.
	1.3 The project and end products should satisfy owner's needs, requirements, and suitable for intended usage.
Principle 2 (Leadership)	2.1 Owner, designer, consultant, contractor are fully responsible for application of quality management system to meet customer requirements and strive to exceed customer expectations by complying with defined scope of work in the contract documents.
	2.2 Every member of the project team should exert collaborative efforts in all the functional areas to improve the quality of the project.

(Continued)

TABLE 1.3 (Continued)
Quality Principles of Construction Projects

Principle	Construction Projects' Quality Principle
Principle 3 (Involvement/ Engagement of People)	3.1 Each member of the project team should participate and fully involved as per their abilities in all the functional areas by adhering to team approach and coordination to continuously improve quality of the project.
Principle 4 (Process Approach)	4.1 Contractor to build the facility as stipulated in the contract documents, plan, specifications as per the approved schedule and within agreed upon budget to meet owner's objectives.
	4.2 Contractor should study all the documents during tendering/bidding stage and submit his/her proposal taking into consideration all the requirements specified in the contract documents and identifying, understanding, and managing interrelated processes as a system in achieving the specified product output.
	4.3 Contractor is responsible to provide all the resources, manpower, material, equipment to build the facility as per specifications to produce the specified products.
	4.4 Contractor to check executed/installed work to confirm that work has been performed/executed as specified, using specified/approved materials, approved shop drawings, installation methods, and specified references, codes, standards to meet intended use.
Principle 5 (System Approach to Management)	5.1 Contractor to prepare contractor's quality control plan (CQCP) and follow the same to ensure meeting the performance standards specified in the contract documents.
	5.2 Method of payments (work progress, material, equipment, etc.) to be clearly defined in the contract documents. Rate analysis of Bill of Quantities (BOQ) or Bill of Materials (BOM) item to be agreed before signing of contract.
	5.3 Contract documents should include a clause to settle the dispute arising during construction stage.
Principle 6 (Continual Improvement)	6.1 Contractor shall follow the submittal procedure specified in the contract documents for detailed design, procurements, checklists, inspection, and testing procedures as per the communication matrix. Review the contents of transmittals and executed work prior to submission for approval.
Principle 7 (Factual Approach to Evidence-Based Design Making)	7.1 Contractor shall follow an agreed-upon quality assurance and quality control plan. Consultant, PMC shall be responsible to oversee the compliance with contract documents and specified standards and codes.
	7.2 Contractor is responsible to construct the facility to produce the products as specified and use the material, products, systems, equipment, and methods that satisfy the specified requirements (factual approach to design making).

(*Continued*)

TABLE 1.3 (Continued)
Quality Principles of Construction Projects

Principle	Construction Projects' Quality Principle
Principle 8 (Mutual Beneficial Relationship) Relationship Management	8.1 Contractor/all team members should participate and put collective efforts to perform the work as per agreed upon construction program and hand over the project as per contracted schedule to meet the owner's requirements.
	8.2 All team members should focus on participative management and strong operational accountability at the individual contributory level to follow principles of total quality management.

Source: Abdul Razzak Rumane. (2021). *Quality Management in Oil and Gas Projects*, reprinted with permission of Taylor & Francis Group.

2 Project Delivery Systems

Abdul Razzak Rumane

2.1 INTRODUCTION

After the establishment of project objectives, the owner/client develops the project procurement strategy by selecting a particular type of project delivery system. The type of project delivery system varies from project to project taking into consideration the objectives of the project. Project delivery method is a system to achieve satisfactory completion of a construction project from the inception to the occupancy.

A project delivery system is defined as the organizational arrangement among various participants comprising the owner, designer, contractor, and many other professionals involved in the design and construction of a project/facility to translate/transform the owner's needs/goals/objectives into a finished facility/project to satisfy the owner's/end user's requirements.

The project delivery system;

- Establishes scope and responsibility for how the project is delivered to the owner.
- Includes project design and construction.
- Defines responsibilities/obligations of each of the participants is expected to perform, such as scheduling, cost control, quality management, safety management, and risk management during various phases of the construction project life cycle (concept design, schematic design, detailed design, construction, and testing, commissioning, and handover).
- Is the approach by which the project is delivered to the owner but is separate and distinct from the contractual arrangements for financial compensation.
- Establishes procedures, actions, and sequence of events to be carried out.

2.2 TYPES OF PROJECT DELIVERY SYSTEMS

There are several types of contract delivery systems. TABLE 2.1 illustrates the most common project delivery systems followed in construction projects.

2.2.1 Design-Bid-Build

In this method, the owner contracts design professional(s) to prepare detailed design and contract documents. These are used to receive competitive bids from the

DOI: 10.1201/9781003636687-2

TABLE 2.1
Categories of Project Delivery Systems

Serial Number	Category	Classification	Sub-Classification
1	Traditional System (Separated and Cooperative)	Design-Bid-Build	Design-Bid-Build
		Variant of Traditional System	Sequential Method
			Accelerated Method
2	Integrated System	Design-Build	Design-Build
		Design-Build	Joint Venture (Architect and Contractor)
		Variant of Design- Build System	Package Deal
		Variant of Design- Build System	Turnkey Method (EPC)—(Engineering, Procurement, Construction)
		Variant of Design-Build System (Turnkey)	Build-Operate-Transfer (BOT)
			Build-Own-Operate-Transfer (BOOT)
			Build-Transfer-Operate (BTO)
			Design-Build-Operate-Maintain (DBOM)
		Variant of Design- Build System (Funding Option)	Lease-Develop-Operate (LDO)
			Wraparound (Public-Private Partnership)
		Variant of Design- Build System	Build-Own-Operate (BOO)
			Buy-Build-Operate (BBO)
3	Management Oriented System	Management Contracting	Project Manager (Program Management)
		Construction Management	Agency Construction Manager
			Construction Manager at Risk
4	Integrated Project Delivery System	Integrated Form of Contract	

contractors. A design build-bid-build contract has well-defined scope of work. This method involved three steps:

1. Preparation of complete detailed design and contract documents for tendering.
2. Receiving bids from pre-qualified contractors.
3. Award of contract to successful bidder.

In this method, two separate contracts are awarded, one to the designer/consultant and one to the contractor. In this type of contract structure, design responsibility is primarily that of the architect or engineer employed by the client and the contractor(s) is primarily responsible for construction only. In most cases, the owner contracts a designer/consultant to supervise the construction process. These types of contracts are lump sum fixed price contracts. Any variation or changes during the construction need prior approval from the owner. Since a complete design is prepared before construction, the owner knows the cost of the project, the time of completion of the project, and the configuration of the project. The client, through the architect or engineer, retains control of the design during construction. This type of contracting system requires considerable time; each step must be completed before starting the next step. TABLE 2.2 illustrates the main aspects, advantages, and disadvantages of the design-bid-build type (traditional type) of project delivery system whereas FIGURE 2.1 illustrates the design-bid-build contractual relationship.

2.2.1.1 Multiple Prime Contracting

Multiple prime contracting is a variation of the design-bid-build type of project delivery system. In this type of contracting system, the owner holds separate contracts with contractors of various disciplines and retains control over the project. This system can result in lower costs to the owner because it avoids the compounded profit and overhead margins that are common to the single contract method. Since the project work is divided into a number of packages, more firms take part in bidding which results in lower prices and also reduces liability for delays by postponing the bidding of follow-on work. With this type of system "fast tracking" of projects is possible. In this type of project delivery system, the owner normally engages a Construction Manager (CM) or firm to manage the project. FIGURE 2.2 illustrates the contractual relationship in a multiple prime contracting system.

2.2.2 Design-Build

In a design-build contract, the owner contracts a firm singularly responsible to design and build the project. In this type of contracting system, the contractor is appointed based on an outline design or design brief to understand the owner's intent of the project. The owner has to clearly define his/her needs, performance requirements, and comprehensive scope of work prior to signing the contract. It is a must that the project definition is understood by the contractor to avoid any conflict in the future as the contractor is responsible for the detailed design and construction of the project. A design-build type of contract is often used to shorten the time required to complete a project. Since the contract with the design-build firm is awarded before starting any design or construction, a cost-plus contract or reimbursable arrangement is normally used instead of a lump-sum, fixed-cost arrangement. This type of contract requires extensive involvement of the owner during the entire life cycle of a project. He/she has to be involved for making decisions during the selection of design alternatives and the monitoring of costs and schedules during construction and therefore the owner has to maintain/hire a team of qualified professionals to perform these

TABLE 2.2
Design-Bid-Build Delivery System

Project Delivery System	Main Aspects	Advantages	Disadvantages
Traditional Contracting System (Design-Bid-Build)	In this method, the owner contracts design professional(s) to prepare detailed design and contract documents. These are used to receive competitive bids from the contractors. A design-bid-build contract has well-defined scope of work. This method involved three steps: 1. Preparation of complete detailed design and contract documents for tendering. 2. Receiving bids from pre-qualified contractors. 3. Award of contract to successful bidder. In this method, two separate contracts are awarded, one to the designer/consultant and one to the contractor. In this type of contract structure, design responsibility is primarily that of the architect or engineer employed by the client and the contractor(s) is primarily responsible for construction only.	• Traditional, well-known delivery system. • Suitable for all types of clients. • Client has control over design process and quality. • Project scope is defined. • Project schedule is known. • Cost certainty at the time of awarding the contract. • Client has direct contractual control over designer (consultant) and contractor. • Low risk for client. • Contractual roles and responsibilities of all parties are clearly defined and known and understood by each participant. • Well-defined relationship among all the parties. • Higher level of competition resulting in low bid cost.	• Long project life cycle. • Client has no control over selection of contractor/ subcontractor. • Client must have the resources and expertise to administer the contracts of designer and contractor. • Higher level of inspection by supervision agency. • Lack of input from constructors may result in design and constructability issues. • Project cost may be higher than estimated by the designer. • If contract documents are unclear, it raises the unexpected costs drastically. • Project likely to be delayed as it may not be possible to complete as per estimated schedule. • Exposure to risk where unreasonable time is set to complete the design phase. • No incentives to the contractor. • Not suitable for projects that are change-sensitive.

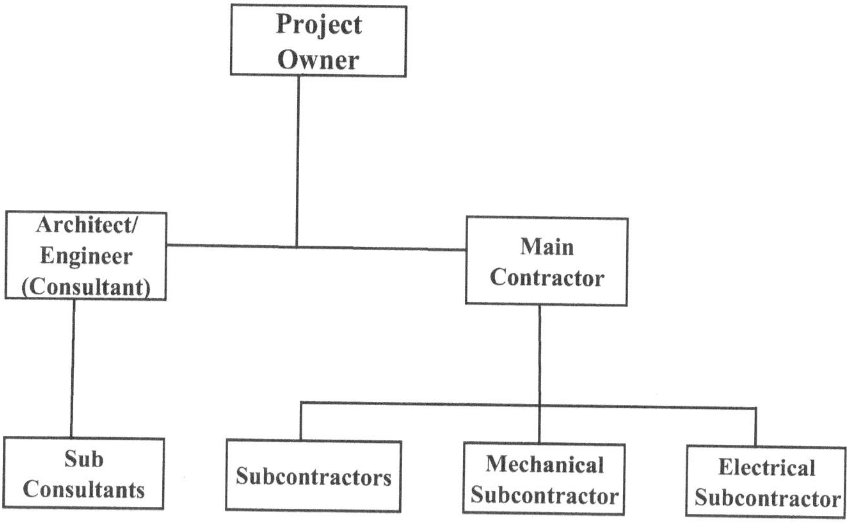

FIGURE 2.1 Design-bid-build contractual relationship.

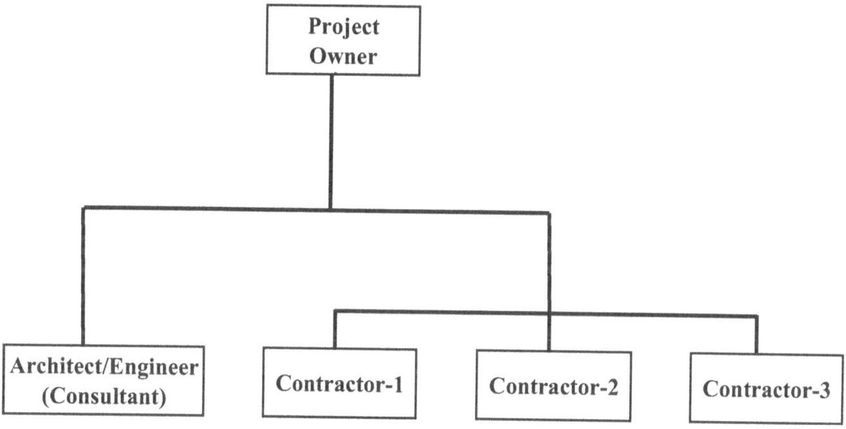

FIGURE 2.2 Multiple prime contractors.

activities. Design-Build contracts are used for relatively straightforward work, where no significant risk or change is anticipated and when the owner is able to specify precisely what is required. TABLE 2.3 illustrates the main aspects, advantages, and disadvantages of the design-build type of project delivery system whereas FIGURE 2.3 illustrates the design-build contractual relationship.

In the case of the design-bid-build type of project delivery system the contractor is contracted after completion of the design based on successful bidding, whereas in the design-build type of deliverable system, the contractor is contracted right from the early stage of the construction project and is responsible for design development of the project.

FIGURE 2.4 illustrates a typical logic flow diagram for the design-bid-build type of construction project and FIGURE 2.5 illustrates a typical logic flow diagram for design-build type of contracting system.

TABLE 2.3
Design-Build Delivery System

Project Delivery System	Main Aspects	Advantages	Disadvantages
Design-Build	In a design-build contract, owner contracts a firm (contractor) singularly responsible for design and construction of the project. In this type of contracting system, the contractor is appointed based on an outline design or design brief to understand the owner's intent of the project. The owner has to prepare a comprehensive scope of work and has to clearly define his/her needs and performance requirements/specifications prior to signing of the contract. It is a must that project definition is understood by the contractor to avoid any conflict in the future as the contractor is responsible for the detailed design and construction of the project. Owner has to be involved for making decisions during the selection of design alternatives and the monitoring of costs and schedules during construction and therefore the owner has to maintain/hire a team of qualified professionals to perform these activities.	• Reduce overall project time because construction begins before completion of design. • Singular responsibility, contractor takes care of the design, schedule, construction services, quality, methods, and technology. • For owner/client, the risk is transferred to design-build contractor. • Project cost defined in early stage and has certainty. • Early involvement of contractor assists constructability. • Suitable for straightforward projects where significant changes or risks are not anticipated and owner is able to precisely specify the objectives/requirements. • Risk management is better than design-bid-build.	• Not suitable for complex projects. • Owner has reduced control over design quality. • Extensive involvement to ensure design deliverable meets project performance requirements. • Extensive involvement of owner during entire life cycle of project. • Real price for a contract cannot be estimated by the owner/client in the beginning. • Changes by owner can be expensive and may result in heavy cost penalties to the owner. • Poor identification of owner needs and wrong understanding of project brief/concept can cause the main problem during the project realization. • Project quality cannot be assured if it is not monitored properly by the owner. • For contractor, more risks in this type of contract. • Not suitable for renovation projects.

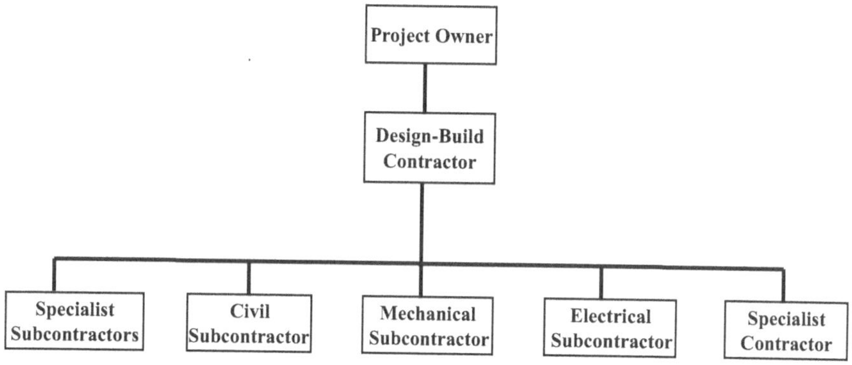

FIGURE 2.3 Design-build contractual relationship.

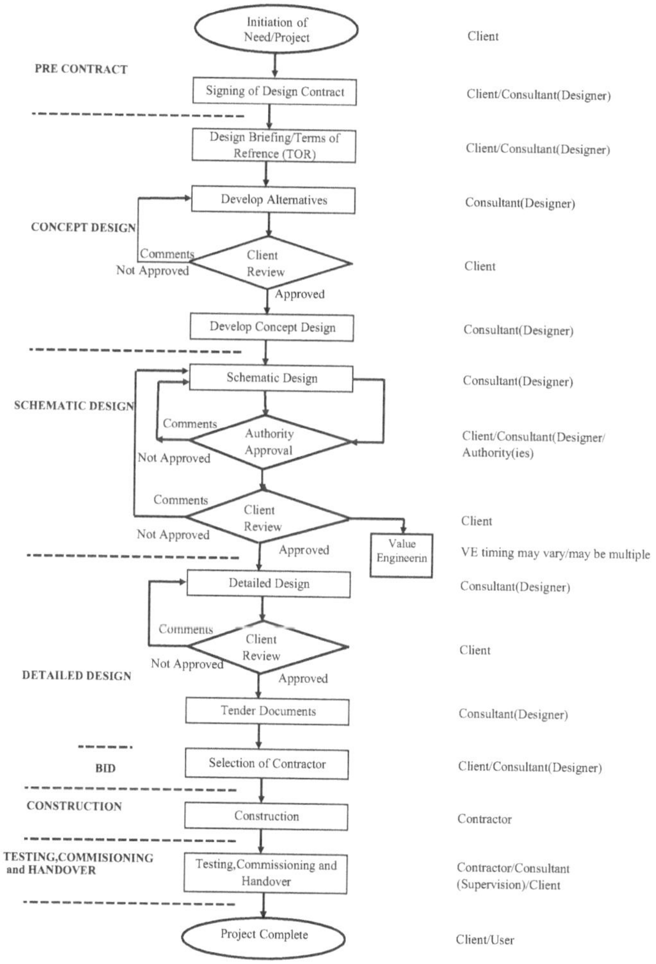

FIGURE 2.4 Logic flow diagram for design-bid-build system.

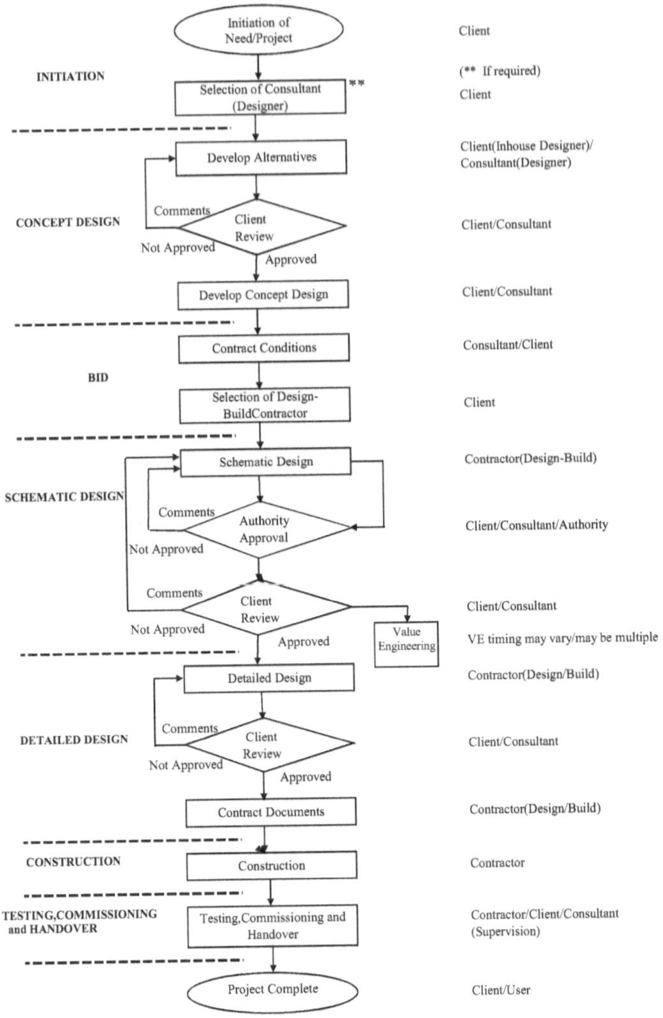

FIGURE 2.5 Logic flow diagram for design-build system.

2.2.3 The Turnkey Contract (EPC)

As the name suggests, these are the types of contracts where on completion, the owner turns a key in the door and everything is working to full operating standards. In this type of method, the owner employs a single firm to undertake design, procurement, construction, and commissioning of the entire work. The firm is also involved in the management of projects during the entire process of the contract. The client is responsible for the preparation of their statement requirements which becomes the strict responsibility of the contractor to deliver. This type of contract is used mainly for the process type of projects and is sometimes called Engineering, Procurement, and Construction (EPC). TABLE 2.4 illustrates the difference between design-build and EPC types of project delivery systems.

TABLE 2.4
Difference between Design-Build and EPC

Design-Build	EPC
• Client/owner has an input into the outline design or design brief to understand owner's intent of the project. • Client/owner contracts a designer/consultant (in case in-house facility not available) to decide design outline. • Detailed design is carried out by the design-build contractor (singular responsibility). • Client/owner has to employ a team of professionals to perform supervision/ management of detailed design, monitor quality, schedule, and cost. • Claims risk is higher. • Mainly suitable for building projects.	• Client has an input in the output (operating capacity). • EPC (engineering-procurement-construction) is a direct contract between client/owner to build a complete project to meet agreed-upon output. • All project activities are carried out by EPC contractor. • Client/Owner is not involved with detailed design processes, except in the event of variation and quality procedures. • Claims risk is lower. • Mainly used for projects such as power plants, process industry, oil, and gas sector.

2.2.4 Build-Own-Operate-Transfer

This type of method is generally used by governments to develop public infrastructure by involving the private sector to finance, design, operate, and manage the facility for a specified period and then transfer the facility to the government free of charge. The terms BOOT and BOT are used synonymously.

Examples of BOT projects include:

- Airports
- Bridges
- Motorways/Toll Roads
- Parking Facilities
- Tunnels

Certain countries allow the private sector to develop commercial and recreational facilities on government land through the BOT scheme.

2.2.5 Project Manager

A project manager contract is used by the owner when the owner decides to turn over the entire project management to a professional project manager. In the project manager type of contract, the project manager is the owner's representative and is directly responsible to the owner. The project manager is responsible for planning, monitoring, and management of the project. In the broadest sense, the project manager has the responsibility for all the phases of the project from the inception of the project till the completion and handing over of the project to the owner/end user.

The project manager is involved in giving advice to the owner and is responsible to appoint design professional(s), consultant, and supervision firm and selecting the contractor to construct the project. TABLE 2.5 illustrates the main aspects, advantages, and disadvantages of the project manager type of project delivery system whereas FIGURE 2.6 illustrates the project manager contractual relationship.

TABLE 2.5
Project Manager Type Delivery System

Project Delivery System	Main Aspects	Advantages	Disadvantages
Project Manager (PM)	A project manager type of delivery system is used by the owner when the owner decides to turn over the entire project management to a professional project manager. In project manager type of contract, the project manager is the owner's representative and is directly responsible to the owner. The project manager acts as a management consultant on behalf of the owner. The project manager is responsible for planning, monitoring, and management of the project. In its broadest sense, the project manager has responsibility for all the phases of the project from inception of the project till the completion and handing over of the project to the owner/end user. Project manager is involved in giving advice to the owner and is responsible to appoint design professional(s), consultant, supervision firm, and select the contractor/package contractor to construct the project.	• Owner/client retains full control of design. • Provide the opportunity for "fast track" or overlapping design and construction phases. • Reduces the owner's general management and oversight responsibilities. • Changes can be accommodated in unlet packages as long as there is no impact on time and cost. • Expert opinion and independent view towards constructability, cost, value engineering and team member selection. • Multi prime type of delivery system is possible.	• Added project management cost to the owner. • Owner must have resources and expertise to deal with PM. • Owner cedes much of day-to-day control over the project to the PM. • Project manager is not at risk to the cost. • Owner continues to hold construction contracts and retains contractual liability.

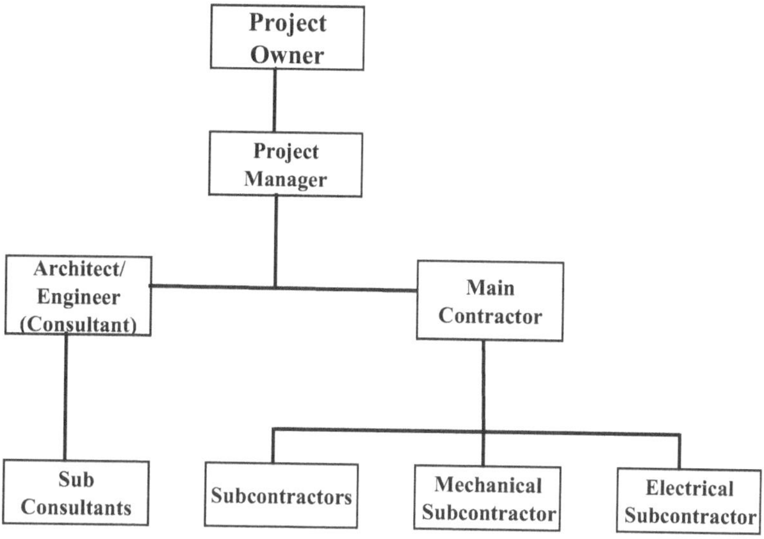

FIGURE 2.6 Project manager contractual relationship.

2.2.6 Construction Management

In this method, the owner contracts a construction management firm to coordinate the project for the owner and provide construction management services. There are two general forms of construction management. These are:

1. Agency Construction Management (Agency CM).
2. Construction Management at Risk (CM at Risk).

The agency construction management type is a management process type of contract system having a four-party arrangement involving owner, designer, construction management firm, and contractor. The construction manager provides advice to the owner regarding cost, time, safety, and about the quality of materials/products/systems to be used on the project. The agency CM firm performs no design or construction but assists the owner in selecting design firm(s), and contractor(s) to build the project.

Agency construction management could be implemented in conjunction with any type of project delivery system.

The basic concept of construction management type of contract is that the firm is knowledgeable and capable of coordinating all aspects of the project to meet the intended use of the project by the owner. The agency construction manager acts as a principal agent to advise the owner/client whereas the construction manager at risk is responsible for on-site performance and actually performs some of the project work. CM at Risk type of contract has two stages. The first stage encompasses pre-construction services and during the second stage, the CM at Risk is responsible for performing the construction work. CM at Risk project delivery system is also known as Construction Manager/General Contractor (CM/GC).

TABLE 2.6 illustrates the main aspects, advantages, and disadvantages of construction management at risk type of project delivery system whereas FIGURE 2.7

TABLE 2.6
Construction Management Delivery System

Project Delivery System	Main Aspects	Advantages	Disadvantages
Construction Management	In this method, owner contracts a construction management firm to coordinate the project for the owner and provide construction management services. There are two general types of construction management type of contracts. These are: 1. Agency construction management. 2. Construction management at risk. The agency construction management is a management process system that can be implemented regardless of project delivery method. Agency CM has a four party arrangement involving owner, designer, construction management firm, and contractor. CM at risk type of delivery system has two contracts, one between owner and designer and other one between owner and the CM at Risk. CM at Risk is selected on qualification basis. The construction manager, a person, or a firm, provides advice to the owner regarding cost, time, safety, and the quality of materials/products/systems to be used on the project. The architect/engineer is responsible for design of the project. The basic concept of construction management type of contract is that the firm is knowledgeable and capable of coordinating all aspects of the project to meet the intended use of the project by the owner. Agency construction manager acts as an advisor to the owner/client whereas construction manager at risk is responsible for on-site performance and actually performs some of the project work while during the design phase construction manager acts as consultant to the owner and offers pre-construction services. The agency CM performs no design or construction but assists the owner in selecting design firm(s), and contractor(s) to build the project.	• Client retains full control of project. • Agency CM helps owner providing advice during the design phase, bid evaluation, overseeing construction, managing project schedule, cost, and quality. • Suitable for large and complex projects with multiple phases and contract packages. • Since construction manager is an expert entity, it provides an independent view regarding constructability, cost, and value engineering. • Shorter project schedule. • Changes can be accommodated in unlet packages.	• Construction manager takes control of packages and interaction with contractors/subcontractors but has no contractual role. • No cost certainty till final package is let. • Client retains all the risk. • Client has to manage the contractual agreements with each package contractor. • Client has to carry majority risk.

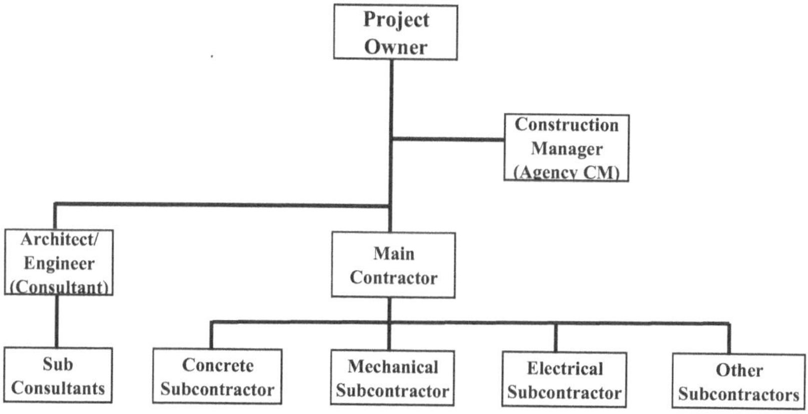

FIGURE 2.7 Construction manager contractual relationship (agency CM).

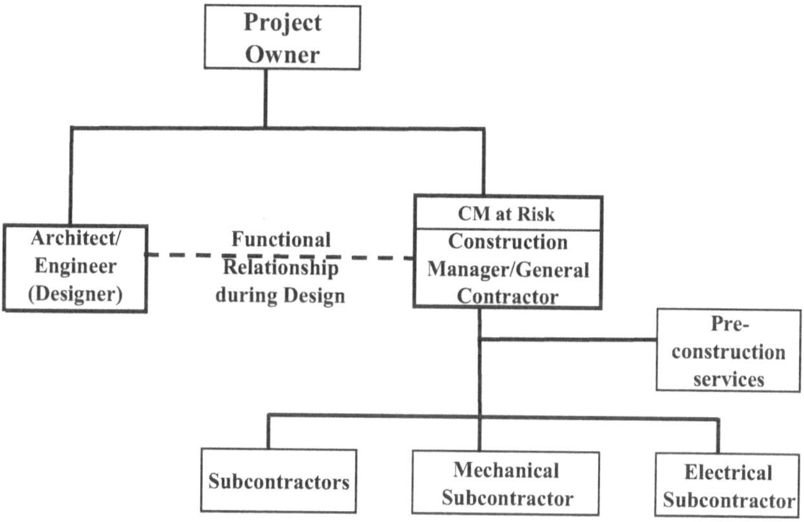

FIGURE 2.8 Construction manager contractual relationship (CM at Risk).

illustrates the contractual relationship for agency CM and FIGURE 2.8 illustrates the contractual relationship for CM at risk.

2.2.7 Integrated Project Delivery

In this delivery method the owner, designer, and contractor are contractually required to collaborate among themselves so that the risk responsibility and liability for the project delivery are collectively managed and appropriately shared. TABLE 2.7 illustrates the main aspects, advantages, and disadvantages of the Integrated Project Delivery (IPD) type of project delivery system and FIGURE 2.9 illustrates the contractual relationship (IPD) between various parties.

TABLE 2.7
Integrated Project Delivery System

Project Delivery System	Main Aspects	Advantages
Integrated Project Delivery System	Integrated project delivery system is an alternative project delivery method distinguished by a contractual agreement at a minimum between owner, design professional, and contractor where the risks and rewards are shared and stakeholder's success is dependent on project. It is a relational contracting that sets performance and results expectations. It creates an integrated team which results in optimization of the project. Value analysis throughout the design process, not at the end. It has full transparency wherein the information is shared openly and only consensus decisions are implemented.	• Integrated form of agreement between owner, architect/engineer, and contractor. • Owner's business plan as the basic design criteria. • Design and cost of the project stay in balance. • Common understanding of project parameters, objectives, and constraints. • Better relationship among project teams. • Better collaboration. • More certainty of outcome within the agreed-upon target cost, schedule, and program. • Superior risk management. • Establishes a collaborative governance structure. • Incorporates Lean project delivery.

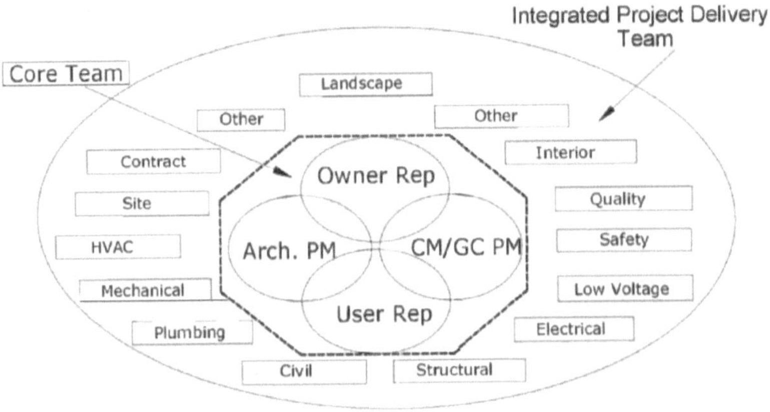

FIGURE 2.9 Integrated project delivery system contractual relationship.

In all types of project delivery systems, there are mainly three participants.

1. Owner
2. Designer (A/E)
3. Contractor (known as main contractor or general contractor)

However, if the owner engages a construction manager or project manager then it becomes a four-party contract. TABLE 2.8 summarizes relationships among various project participants.

2.3 TYPES OF CONTRACTS/PRICING

A contract is a formalized means of agreement, enforceable by law, between two or more parties to perform certain work or provide certain services against agreed-upon financial incentives to complete the works/services. Regardless of the type of project delivery system selected, the contractual arrangement by which the parties are compensated must also be established. In construction projects, determining how to procure the product is as important as determining what and when. In most procurement activities there are several options available for purchase, or subcontracts. The basis of compensation type relates to the financial arrangement among the parties as to how the designer or contractor is to be compensated for their services. The following are the most common types of contract/compensation methods:

1. Firm Fixed Price or Lump Sum Contract.
   a. Firm fixed price.
   b. Fixed price incentive fee.
   c. Fixed price with economic adjustment price.

TABLE 2.9 illustrates the advantages and disadvantages of fixed-price or lump-sum price contracts.

2. Unit Price.
   a. Specific price for a particular task or unit of work performed by the contractor.
   b. In this method each unit of work to be precisely defined.

TABLE 2.10 illustrates the advantages and disadvantages of unit price contracts.

3. Cost Reimbursement Contract (Cost Plus).
   a. Cost plus percentage fee.
   b. Cost plus fixed fee.
   c. Cost plus incentive fee.
   d. Cost plus award fee.

TABLE 2.11 illustrates the advantages and disadvantages of cost reimbursement contracts.

TABLE 2.8
Relationship among Project Participants

Serial Number	Category	Classification	Sub-Classification	Owner/ Designer	Owner/ Contractor	Designer/ Contractor	Owner/ CM	CM/ Designer	CM/ Contractor
1	Traditional System	Design-Bid-Build	Design-Bid-Build	C	C	F	–	–	–
2	Integrated System	Design-Build	Design-Build	C	C	I	–	–	–
				Contractual Relationship					
3	Management Oriented System	Construction Management	Agency Construction Manager	C	C	–	C	F	F
			Construction Manager at Risk	C	–	C	C	F	C

Legend: C-Contractual Relationship, F-Functional Relationship, I-Integral Part of Contract

TABLE 2.9
Fixed or Lump Sum Price Contracts

Serial Number	Project Contracting Type	Main Aspects	Advantages	Disadvantages
1	Fixed Price or Lump Sum	With this type of contract, the contractor agrees to perform the specified work/services with a fixed price (it is also called lump sum) for the specified and contracted work. Any extra work is executed only upon receipt of instruction from the owner. Fixed price contracts are generally inappropriate for work involving major uncertainties, such as work involving new technologies.	• Low financial risk to the owner. • Total cost of project known before construction. • Project viability is known before a commitment is made. • Suitable for projects that can be completely designed and whose quantities are definable. • Minimal owner supervision. • Contractor will usually assign its best personnel to the work. • Maximum financial motivation of contractor. • Contractor has to solve its own problems and do so quickly. • Contractor selection by competitive bidding is fairly easy, apart from the deliberately low price.	• High financial risk to the contractor. • Owner bears the risk of poor quality from the contractor trying to maximize the profit within fixed cost. • Contractor's price may include high contingency. • Variation can be time consuming and costly. • Variations (changes) are difficult and costly. • More time taken for bidding and for developing a good design basis. • Contractor will tend to choose the cheapest and quickest solutions, making technical monitoring, and strict quality control by the owner essential.

TABLE 2.10
Unit Price Contracts

Serial Number	Project Contracting Type	Main Aspects	Advantages	Disadvantages
1	Unit Price	The owner and contractor agree to structure the contract on specified unit price for the estimated quantities of the work. In unit price contract, the work to be performed is broken into various parts and a fixed price is established for each unit of work. This type of contract is well-suited for repetitive or easily quantifiable tasks. A unit price contract provides benefits to both owners and contractors.	• Complete design definition is not essential for the bidding process. • Suitable for competitive bidding and relatively easy contractor selection subject to sensitivity evaluation. • Typical drawings can be used for bidding. • Bidding is speedy and inexpensive, and an early start is possible. • Flexibility—depending on the contract conditions, the scope and quantity of work can be varied and easily adjustable.	• The final cost is not known at the outset. • Additional site staff are needed to measure, control, and report on the cost and the status of the work. • Biased bidding and front-end loading may not be detected.

TABLE 2.11
Cost Reimbursement Contracts

Serial Number	Project Contracting Type	Main Aspects	Advantages	Disadvantages
1	Cost Reimbursement Contract (Cost Plus)	It is a type of contract in which contractor agrees to do the work for the cost related to the project plus an agreed upon amount of fee that covers profits and non-reimbursable overhead costs. Following are different types of cost-plus contracts: 1. Cost-plus % fee contract. 2. Cost-plus fixed fee contract. 3. Cost-plus incentive fee contract.	• Suitable to start construction concurrently while design is in progress. • Suitable for renovation type of projects. • Suitable for projects expecting major changes. • Suitable for projects with possible introduction of latest technology. • Flexibility in dealing with changes. • An early start can be made. • Useful where site problems such as trade union actions like delays or disruptions may be encountered. • Owner can control all aspects of the work.	• Possibility of overspending by the contractor. • Difficult to predict final cost. • Project quality is likely to be affected as the fee will be same no matter how low the costs are. • Final cost is unknown. • Difficulties in evaluating proposals, for example, L1 may not result in selecting a contractor in achieving lowest project cost. • Contractor has little incentive for early completion or cost economy. • Contractor may assign its "second division" personnel, make excessive use of agency personnel, or use the job as a training vehicle for new personnel. • Biased bidding of fixed fees and reimbursable rates may not be detected.

TABLE 2.12
Re-Measurement Contracts

Serial Number	Project Contracting Type	Main Aspects	Advantages	Disadvantages
1	Re-Measurement Contract	It is a contract in which contractor is paid as per the actual quantities of the work done. In this type of contract in which contractor agrees to do the work based on one of the following criteria: 1. Bill of quantities, or 2. Schedule of rates, or 3. Bill of material. In this type of contract, payment is linked to measured work completion.	• Suitable for competitive bidding. • Fair competition. • For contractor, low risk. • Suitable for projects where quantities of work cannot be determined in advance of construction. • Fair basis for competition.	• Require adequate breakdown and design definition of work unit. • Final cost not known with certainty until the project is complete. • Additional requirements of administrative staff to measure, control, and report.

4. Re-Measurement Contract.
   a. Bill of quantities.
   b. Schedule of rates.
   c. Bill of materials.

TABLE 2.12 illustrates the advantages and disadvantages of re-measurement contract.

5. Target Price Contract.

TABLE 2.13 illustrates the advantages and disadvantages of target price contract.

6. Time and Material Contract.

TABLE 2.14 illustrates the advantages and disadvantages of time and material contracts.

7. Guaranteed Maximum Price (GMP).
   a. Cost plus fixed fee Guaranteed Maximum Price (GMP) contract.
   b. Cost plus fixed fee Guaranteed Maximum Price (GMP) and bonus contract.
   c. Cost plus fixed fee Guaranteed Maximum Price (GMP) with arrangement for sharing any cost-saving type contract.

TABLE 2.13
Target Price Contracts

Serial Number	Project Contracting Type	Main Aspects	Advantages	Disadvantages
1	Target Price Contract	A target cost contract is based on the concept of top-down approach, which provides a fixed price for an agreed range of out-turn costs around the target. In this type of contract overrun or underspend are shared by the owner and the contractor at a predetermined agreed-upon percentage.	• Final cost is known. • Contractor may share the savings as per agreed-upon percentages. • Flexibility in controlling the work. • Almost an immediate start on the work, even without a scope definition. • Encourages economic and speedy completion (up to a point). • Contractor is rewarded for superior performance.	• Tight cost control is required. • Difficult to adjust major variations or cost inflation. • No opportunity to competitively bid the targets. • Difficulty in agreeing on an effective target for superior performance. • Variations are difficult and costly once the target has been established. • If the contractor fails to achieve the targets, it may attempt to prove the owner's fault.

TABLE 2.14
Time and Material Contracts

Serial Number	Project Contracting Type	Main Aspects	Advantages	Disadvantages
1	Time and Material Contract	A time and material contract has elements of unit-price and cost-plus type of contract. This type of contract is mainly used for maintenance contracts and small projects.	• Owner pays the contractor based on actual cost for material and time as per pre-agreed rates. • Suitable for design services where it is difficult to determine total expected efforts in advance.	• Contractor must be accurate in estimating the price, which normally includes indirect and overhead costs.

2.3.1 Guaranteed Maximum Price

In this method, the contractor is compensated for actual costs incurred in connection with the design and construction of the project, plus a fixed fee all subject, however, to a ceiling above which the client is not obligated to pay. It guarantees maximum price contract price specifies a target profit (or fee), a price ceiling (but not for profit ceiling or floor), and a profit (or fee) adjustment formula. These elements are all negotiated at the outset. Guaranteed maximum price combines construction management with design-build. With a guaranteed maximum price contract, amounts below the maximum are typically shared between the client and the contractor, while the contractor is responsible for absorbing the cost above the maximum. Any changes that result from the specific instructions of the owner fall outside the guaranteed price. Cost-plus Guaranteed maximum price, as it is also known, type of contract is typically used:

- Where time pressure requires letting of the contract before design development is sufficiently advanced to allow a conventional lump-sum type of contract to be fixed.
- If this type of contract is likely to be less costly than other types.
- Where financing or other constraints preclude the use of alternatives such as two-stage contract of construction management.
- If it is impractical to obtain a certain type of service with improved delivery, technical performance, or quality without the use of this type of contract.

In this method, the contractor and owner have the knowledge that the drawings and specifications are not complete, and the contractor and the owner agree to work together to complete the drawings and specifications as provided in the contract agreement. This type of contract is weighed heavily in favor of the owner. The contractor takes on all the risks in this type of contracting system. Value engineering studies are conducted to identify design alternatives to help projects maintain the budget and schedule. This type of contract need:

- Adequate cost pricing information for establishing a reasonable firm target price at the time of initial contract negotiation.
- Contractor's tendering/bidding department should have adequate information to provide necessary data to support negotiation of final cost and incentive price revision.
- High administration costs from owner side to monitor what the contractor is actually spending to get the benefit of under spending.
- Evaluation of a minimum of two or three proposals for any major subcontract work.

In certain GMP contracts, the owner monitors and controls the contractor's expenses towards the project resources such are construction equipment, machinery, manpower, and staff on a monthly basis by fixing the basic price and profit percentage agreed upon at the initial stage.

TABLE 2.15 illustrates the advantages and disadvantages of GMP-type contracts.

2.4 SELECTION OF PROJECT DELIVERY SYSTEM

Each of the project delivery systems previously discussed has advantages and disadvantages. It is a strategic decision that the owner has to make considering the suitability of the appropriate system to achieve successful completion of the project. Therefore, while selecting an appropriate project delivery system the owner has to consider following:

- Size and complexity of the project.
- Type of project.
- Location of project.
- Owner's level of construction expertise (human resources available with owner and owner's knowledge of construction management practices).
- Owner's interest to exert influence/control over the design.
- Owner's interest to exert influence/control over the management of planning.
- Owner's interest to exert influence/control over the management of construction.
- Owner's interest to exert influence/control over the management of project and the end user(s).
- Design.
- Schedule.
- Budget, funding mechanism.
- Quality.
- Risk owner can tolerate (risk allocation).

TABLE 2.15
Guaranteed Maximum Price Contracts

Serial Number	Project Contracting Type	Main Aspects	Advantages	Disadvantages
1	Cost Plus Guaranteed Maximum Price Contract	With this type of contract, the owner and the contractor agree to a project cost guaranteed by the contractor as maximum. In this method, the contractor is compensated for actual costs incurred, in connection with design and construction of the project, plus a fixed fee-all subject, however, to a ceiling above which the client is not obligated to pay. This type of contract needs adequate cost pricing information for establishing a reasonable firm target price at the time of initial contract negotiation.	• Maximum contract price is certain. • Construction can start at an early stage.	• Contractor shares the maximum risk. • High administration cost from owner side to monitor what the contractor is actually spending to get the benefit of under spending. • Contractor's tendering/bidding department should have adequate information to provide necessary data to support negotiation of final cost and incentive price revision.

2.4.1 Selection of Project Delivery Teams (Designer/Consultant, Contractor)

Construction projects have the involvement of three main parties. These are:

1. Owner
2. Designer/Consultant
3. Contractor

Participation involvement of all three parties at different construction phases is required to develop a successful facility/project. These parties are involved at different levels and their relationship and responsibilities depend on the type of project delivery system and contracting system. The following are the common procurement methods for selection of project teams:

1. Low Bid.
    - Selection is based solely on the price.
2. Best Value.
    a. Total Cost
        - Selection is based on total construction cost and other factors.
    b. Fees
        - Selection is based on a weighted combination of fees and qualifications.
3. Qualification Based Selection (QBS).
    - Selection is based solely on qualification.

TABLE 2.16 illustrates the main aspects of selection methods and selection criteria of the project delivery system team.

2.4.1.1 Selection of Designer/Consultant

Normally, design professionals (consultants) are hired on the basis of **Qualifications**. The qualification-based selection can be considered as meeting one of the 14 points of Deming's principles of transformation which states "End the practice of awarding business on the basis of price alone." The basis of selection is solely based on demonstrated competence, professional qualification, and experience for the types of services required. In quality-based selection (QBS) the contract price is negotiated after the selection of the best-qualified firm. TABLE 2.17 lists the criteria for the selection of construction project designer/consultant on a QBS basis.

Upon selection of designer/consultant by the client, the designer is invited to submit the proposal for design services.

FIGURE 2.10 illustrates the procedure for submission of design services proposal and FIGURE 2.11 illustrates the overall scope of work normally performed by the designer /consultant.

2.4.1.1.1 Request for Proposal (RFP)

It is a project-based process involving solutions, qualifications, and price as the main criteria that define a winning proponent. It is a solicitation document requesting proponents request proposal in response to the required scope of services. The

TABLE 2.16
Procurement Selection Type and Selection Criteria for Project Delivery System

Serial Number	Selection Type	Description	Selection Criteria D-B-B	D-B	CM at Risk	IPD
1	Low Bid	Total construction cost including the cost of work is the sole criterion for final selection.	Yes	Yes	Not Typical	Not Typical
2	Best Value					
	2.1 Total cost	Both total cost and other factors are criteria in the final selection.	Yes	Yes	Not Typical	Not Typical
	2.2 Fees	Both fees and qualifications are factors in the final selection.	Not Applicable	Yes	Yes	Yes
3	Qualification Based Selection (QBS)	Total costs of the work are **not** a factor in the final selection. Qualification is the sole factor used in the final selection.	Not Applicable	Yes	Yes	Yes

TABLE 2.17
Request for Qualification (Prequalification of Contractor)

Serial Number	Items to be Evaluated
1	Organization's registration and license
2	Management plan and technical capability
3	Quality certification and quality management system
4	LEED or similar certification
5	Number of awards
6	Design capacity to perform the work
7	Financial strength and bonding capacity
8	Professional indemnity
9	Current load
10	Experience and past performance in similar types of work
11	Experience and past performance of proposed individuals in similar projects
12	Professional certification of proposed individuals
13	Experience and past performance with desired project delivery system

(*Continued*)

TABLE 2.17 (Continued)
Request for Qualification (Prequalification of Contractor)

Serial Number	Items to be Evaluated
14	Design of similar value projects in past
15	List of successfully completed projects
16	Proposed design approach in terms of: a. Performance b. Effectiveness c. Maintenance d. Logistic support e. Environment f. Green building
17	Design team composition
18	Record of professionals in timely completion
19	Safety consideration in design
20	Litigation
21	Price schedule

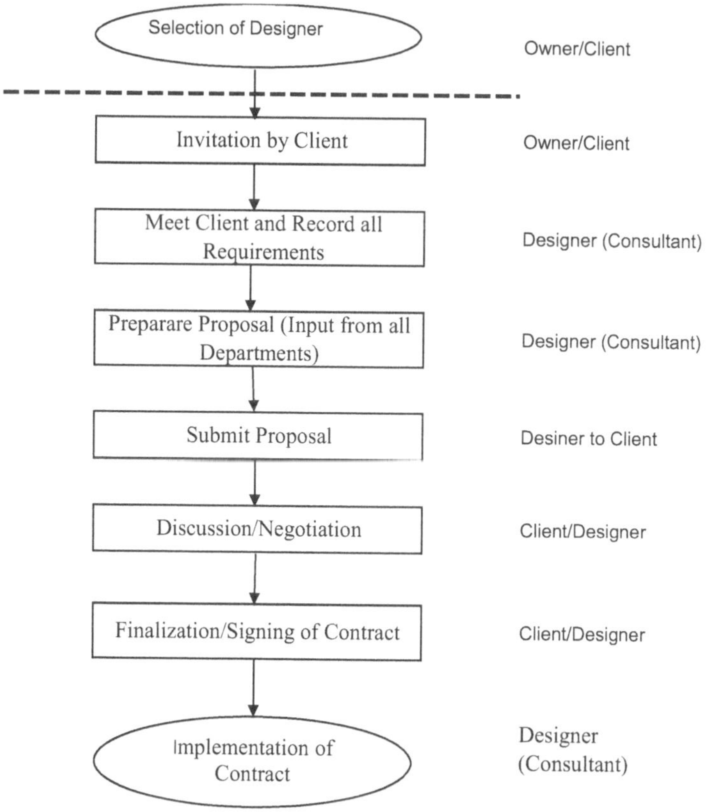

FIGURE 2.10 Contract proposal procedure for designer.

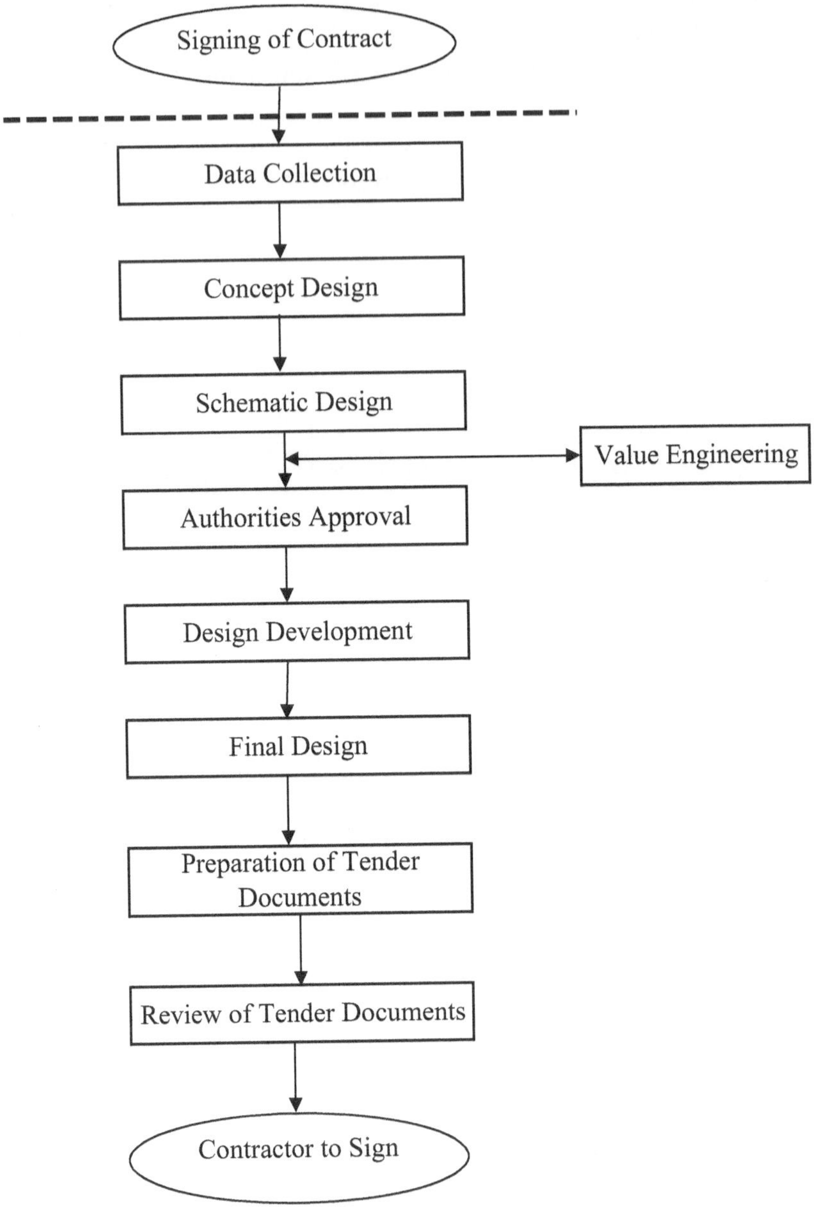

FIGURE 2.11 Overall scope of work of designer.

document does not specify in detail how to accomplish or perform the required services. Request for Proposal can range from a single-step process for straightforward procurement to a multistage process for complex and significant procurement. TABLE 2.18 illustrates the requirements to be included in the Request for Proposal (RFP) document for construction project designer/consultant.

TABLE 2.18
Qualification Based Selection of Architect/Engineer (Consultant)

Serial Number	Content
Project Details (Project Objectives)	
1	Introduction
2	Project description
3	Project delivery system
4	Designer's/consultant's scope of work
5	Preliminary project schedule
6	Preliminary cost of project
7	Type of project delivery system
Sample Questions (Information for Evaluation)	
1	Consultant Name
2	Address
3	Quality management system certification
4	Organization details
5	Type of firm such as partnership or limited company
6	Is the firm listed on stock exchange
7	List of awards, if any
8	Design production capacity
9	Current workload
10	Insurance and bonding
11	Experience and expertise
12	Project control system
13	Design submission procedure
14	Design review system
15	Design management plan
16	Design methodology
17	Submission of alternate concept
18	Quality management during design phase
19	Design firm's organizational chart: a. Responsibility matrix b. CVs of design team members
20	Designer's experience with green building standards or highly sustainable projects
21	Conducting value engineering
22	Authorities' approval
23	Data collection during design phase
24	Design responsibility/professional indemnity
25	Designer's relationship during construction
26	Preparation of tender documents/contract documents
27	Review of tender documents
28	Evaluation process and criteria
29	Any pending litigation
30	Price schedule

2.4.1.2 Selection of Contractor

In construction projects, the selection process for contractors can range from simply deciding to directly award a contract to a multistage process that involves information gathering about the contractor through Request for Information (RFI), prequalification questionnaire (Request for Qualification (RFQ)), and soliciting activities. Normally every owner maintains a list of prospective and previously qualified sellers. The existing list of potential contractors can be expanded by placing advertisements in general publications such as newspapers, trade publications, or professional journals. An Internet search can also help generate a list of prospective sellers.

2.4.1.2.1 Request for Information (RFI)

RFI is a procedure where potential contractors are provided with a general or preliminary description of the problem or need and are requested to provide information or advice about how to better define the problem or need or alternative solutions. It may be used to assist in preparing a solicitation document.

2.4.1.2.2 Request for Qualification (RFQ)/Prequalification Questionnaire (PQQ)

RFQ/PQQ is a process that enables the ability to prequalify contractors for a particular requirement and avoid having to struggle with a large number of lengthy proposals. It is a solicitation document requesting proponents to submit their qualifications and special expertise in response to the required scope of services. This process is used to shortlist qualified proponents for the procurement process. TABLE 2.19

TABLE 2.19
Contents of Request for Proposal (RFP) for Designer/Consultant

Serial Number	Element	Description
1	General Information	a. Company name
		b. Full address
		c. Registration details/Business permit
		d. Management details
		e. Nature of company such as partnership, share holding
		f. Stock exchange listing, if any
		g. Affiliated/Group of companies, if any
		h. Membership of professional trade associations, if any
		i. Award winning project, if any
		j. Quality management certification

(Continued)

TABLE 2.19 (Continued)
Contents of Request for Proposal (RFP) for Designer/Consultant

Serial Number	Element	Description
2	Financial Information	a. Yearly turnover b. Current workload c. Audited financial report d. Tax clearance details e. Bank overdraft/letter of credit capacity f. Performance bonding capacity g. Insurance limit
3	Organization Details (General)	a. Core area of business b. How long in the same field of operation c. Quality control/assurance organization d. Grade/classification, if any
	Organization Details (Experience)	a. Number of years in the same business b. Technical capability: i. Engineering ii. Shop drawing production c. List of previous contracts: i. Name of project ii. Value of each contract iii. Contract period of each contract iv. Contract completion delay d. List of failed/uncompleted contracts e. Overall tender success rate f. Claims/dispute/litigation
4	Resources	a. Human resources: i. Management ii. Engineering staff iii. CAD technicians iv. List of key project personnel v. Skilled labors—permanent/temporary vi. Unskilled laborers—permanent/contracted b. List of equipment, machinery c. Human resource development plan
5	Health, Safety, and Environmental	a. Medical facility b. Safety record c. Environmental awareness
6	Project Reference	
7	Bank Reference	

illustrates the request for qualification or the prequalification questionnaire to select a contractor.

2.4.1.2.3 Request for Proposal (RFP)

It is a project-based process involving solutions, qualifications, and price as the main criteria that define a winning proponent. It is a solicitation document requesting proponents request proposal in response to the required scope of services. The document does not specify in detail how to accomplish or perform the required services. Request for proposal can range from a single-step process for straightforward procurement to a multistage process for complex and significant procurement.

2.4.1.2.4 Request for Quotation (RFQ)

RFQ is a priced-based bidding process that is used when complete documents consisting of defined project deliverables, solutions, specifications, performance standards, and schedules are known. Potential bidders are provided with all the related information (documents)—except price—and are requested to submit the price and the evaluation of the bids are only on **price** subject to fulfilling all the required information. Most construction contractors are selected on **price** basis.

TABLE 2.20 illustrates the difference between Request for Proposal (RFP) and Request for Qualification (RFQ).

In construction projects, the contractor's bids and proposals depend on the type of delivery system and contracting methods. TABLE 2.21 is an example evaluation criterion for shortlisting/selection of contractors.

TABLE 2.20
Difference between RFP and RFQ

Request for Proposal	Request for Qualification
• Solicitation document requesting proposal in response to required scope of services. • Does not specify in detail how to accomplish or perform the required services. • The process is used when the services are not well defined. • RFP is normally used for qualification-based selection (QBS). • The document includes the statement that QBS will be used.	• Solicitation document requesting submittal of qualifications or expertise in response to required scope of services. • Specify specific processes, critical steps, and approaches to accomplish the required services. • The process is used to shortlist qualified proponents (Contractors, Professionals) for use in future projects. • Selection is based on the ability of the firm to achieve project objectives effectively and efficiently.

TABLE 2.21
Contractor Selection Criteria

Evaluation Criteria			Weightage	Key Points for Consideration	Review Result
1. General Information					
1	Company Information			Company's current position—a MUST information	**Yes OR No**
2. Financial			**25%**		
1	Total Turn Over (last five years)		25%	Sum of the turnover for the last five years	
2	Values of Current Work-In-Hand		25%	Project value/value of current work-in-hand	
3	Audit Financial Reports		10%	To confirm the ratio given in point three	
4	Financial Standing		30%		
	33%	Assets		Current assets/current liabilities	
	34%	Liabilities		Total liabilities—total equity/total assets	
	33%	Profit/Loss		Net profit before tax/total equity	
5	Bonding and Insurance Limit		10%		
	60%	a. Performance and Bonding Capacity		Provided or not provided	
	40%	b. Insurance		Provided or not provided	

(Continued)

TABLE 2.21 (Continued)
Contractor Selection Criteria

Evaluation Criteria			Weightage	Key Points for Consideration	Review Result
3. Organization Details					
3a. Business			**20%**		
1	Company's Core Area of Business		30%	Degree of satisfactory answers	
2	Experience of Years in Business		30%	No. of years	
3	ISO Certification		15%	Yes or no	
4	Registration/Classification Status		15%	Grade or classification	
5	Organizational Chart		10%	Key staff indicated (name/title), balanced resources, departmental (specialization) diversity, lines of communication	
6	Dispute/Claims			Degree of satisfactory answers	
3b. Experience			**30%**		
	50%	a. Projects' Value		No. of projects with comparable value	
	50%	b. Projects' Type (similar type and complexity)		No. of projects with similar complexity	
4. Resources			**20%**		
1	Personnel		60%		

(Continued)

TABLE 2.21 (Continued)
Contractor Selection Criteria

Evaluation Criteria			Weightage	Key Points for Consideration	Review Result
	30%	Management		No. of managerial staff	
	30%	Engineers		No. of engineers and project staff	
	30%	Technicians		No. of CAD technicians and foreman	
	10%	Staff Development		% turnover spent on training	
2	Technology		10%	% of turnover spent on acquiring latest construction technology	
3	Plant and Equipment		30%	List of plant and equipment	
5. GENERAL			**5%**		
1	Bank References		30%	Provided or not provided	
2	Project References		30%	Provided or not provided	
2	Health, Safety, and Environmental Narration		40%	Degree of satisfactory answers	

3 Construction Management Delivery System

Abdul Razzak Rumane

3.1 INTRODUCTION

Construction is translating the owner's goals and objectives, by the contractor, to build the facility as stipulated in the contract documents, plans, and specifications within budget and on schedule.

Construction projects are mainly capital investment projects. They are customized and non-repetitive in nature. Construction projects are of constantly increasing technological complexity. Construction projects have become more complex and technical, and the relationships and the contractual grouping of those who are involved are also more complex and contractually varied. In addition, the requirements of construction clients are on the increase and, as a result, construction projects (buildings and other facilities) must meet varied performance standards (climate, rate of deterioration, maintenance, and so on). Therefore, to ensure the adequacy of the client brief which addresses numerous complex client/user needs, it is now needed to evaluate the requirements in terms of activities and their interrelationship. The products used in construction projects are expensive, complex, immovable, and long-lived. Generally, a construction project is composed of building materials (civil), electro-mechanical items, finishing items, and equipment. These are normally produced by other construction-related industries/manufacturers. These industries produce products as per their own quality management practices complying with certain quality standards or against specific requirements for a particular project. The owner of the construction project or his/her representative has no direct control over these companies unless he/his representative/appointed contractor commits to buying their product for use in their facility. These organizations may have their own quality management program. In manufacturing or service industries the quality management of all in-house manufactured products is performed by the manufacturer's own team or under the control of the same organization having jurisdiction over their manufacturing plants at different locations.

Construction projects are comprised of a cross section of many different participants. These participants are both influenced by and depend on each other in addition to "other players" involved in the construction process.

DOI: 10.1201/9781003636687-3

Traditional construction projects have the involvement of three main groups. These are:

1. Owner—A person or an organization that initiates and sanctions a project. He/she requests the needs of the facility and is responsible to arrange the financial resources for the creation of the facility.
2. Designer (Architect/Engineer)—This consists of architects, engineers, or consultants. They are the owner's appointed entity accountable to convert the owner's conception and need into a specific facility (project) with detailed directions through drawings and specifications within the economic objectives. They are responsible for the design of the project and in certain cases supervision of the construction process.
3. Contractor—A construction firm engaged by the owner to perform and complete the construction of a specific facility (project) by providing the necessary staff, workforce, resources, materials, equipment, tools, and other accessories to the satisfaction of the owner/end user in compliance with the requirements of contract documents. The contractor is responsible for implementing the project activities and to achieve the owner's objectives.

Construction projects are executed based on a predetermined set of goals and objectives. In order to process the construction project in an effective and efficient manner and to improve control and planning, construction projects are divided into various phases. In traditional thinking, there are five phases of a construction project lifecycle which have further decomposition into various activities. These are Conceptual Design, Preliminary Design (Schematic Design), Detail Engineering (Design Development), Construction and Testing, Commissioning and Handing Over.

Participation involvement of all three parties at different levels of construction phases is required to ensure the completion of construction and make the project most qualitative, competitive, and economical. Construction projects involve coordinated actions and input from many professionals and specialists to achieve defined objectives. There are several types of project delivery systems, contracting systems in which these parties are involved at different levels. All these contract deliverable systems follow generic life cycle phases of construction projects; however, the involvement/participation of various parties differ depending on the type of deliverable system adapted for a particular project.

Complex and major construction projects have many challenges such as delays, changes, disputes, accidents at the site, etc., and therefore need efficient management of the project from the beginning to the end of construction of the facility/project to meet the intended use and owner's expectations. The main area of construction management covers planning, organizing, executing, and controlling to ensure that the project is built as per the defined scope, maintaining the completion schedule, and within the agreed-upon budget. The owner/client may not have the necessary staff/

resources in-house to manage the planning, design, and construction of the construction project to achieve the desired results. Therefore, in such cases, the owners engage a professional firm or a person, called a construction manager, who is trained and has expertise in the management of construction processes, to assist in developing bid documents, overseeing, monitoring, controlling, and coordinating projects for the owner. The basic construction management concept is that the owner assigns the contract to a firm or a person who is knowledgeable and capable of coordinating all the aspects of the project to meet the intended use of the project by the owner. In the construction management type of project delivery system, the consultant (architect/engineer) prepares complete design drawings and contract documents, then the project is put in for competitive bid and the contract is awarded to the competitive bidder (contractor). The owner hires a third party (construction manager) to oversee and coordinate the construction. The professionals and specialists engaged by the owner as CM bring knowledge and experience that contribute to decisions at every stage of the project for successful completion.

3.2 PROJECT MANAGEMENT AND CONSTRUCTION MANAGEMENT

As per PMI PMBOK, 7th edition, project management is the application of knowledge, skills, tools, and techniques to project activities to meet the project requirements. Project management totally deals with five process groups (Initiating, Planning, Executing, Monitoring and Control, and Closing) divided into 13 knowledge areas that are spread over 47 locally grouped project management processes. Project management involves managing all types of projects from start to finish.

Construction management is a professional management practice applied effectively to the construction project from the inception to the completion of the project for the purpose of managing (planning, organizing, executing, and controlling) schedule, cost, scope, and quality. Construction management services are generally offered by registered engineering firms/professionals having the ability and expertise to manage construction projects.

Construction management is a discipline and management system specially tailored to promote the successful execution of capital and complex projects.

Construction management mainly involves construction-related management activities such as:

- Planning
- Scheduling
- Monitoring and Control
- Quality Control/Quality Assurance
- Human Resources
- Material, Equipment
- Safety and Environmental Protection

TABLE 3.1
Difference between Project Management and Construction Management

Project Management	Construction Management
• Project management is a conventional method of construction administration. • Client/owner retains full control of design and construction. • Client/owner retains direct control over all aspects and quality of the project. • Client/owner has direct contract with the designer (consultant) and contractor, subcontractors. • Client/owner has direct contractual relationship with both the architect/engineer (consultant) and contractor. • Client/owner selects the consultant and contractor.	• Construction management is a progressive more convenient method of construction project administration. • There are two forms of construction management. 1. Agency CM 2. CM at risk • In the case of agency CM type of management system the client/owner has three contracts. One between the owner and designer, one between the owner and contractor, and one between owner and construction manager. • In the case of at-risk CM type of project delivery system the client/owner has two contracts, one between owner and designer and one between owner and at-risk CM/general contractor. • In the agency CM type of management system the CM is not at risk for the budget, schedule, or the performance of the work. • In the at-risk CM type of project delivery system construction services contracts between GC and contractors of different trades. • In the at-risk CM type of project delivery system the CM is responsible for performing the construction and carries financial liability to complete the project within schedule and budget.

The construction manager is responsible for overseeing the performance of contractor(s) towards construction-related activities (engineering design, construction process, testing, commissioning, and handover).

Construction management and project management are different types of project delivery systems. TABLE 3.1 illustrates the basic differences between project management and construction management types of systems.

3.3 ROADMAP FOR CONSTRUCTION MANAGEMENT

FIGURE 3.1 represents a pictorial roadmap that covers all the activities that are discussed in Sections 3.3.1 through 3.3.5.

3.3.1 Project Initiation

Most construction projects begin with the recognition of a new facility. The owner of the facility could be an individual, a public/private sector company, or a governmental

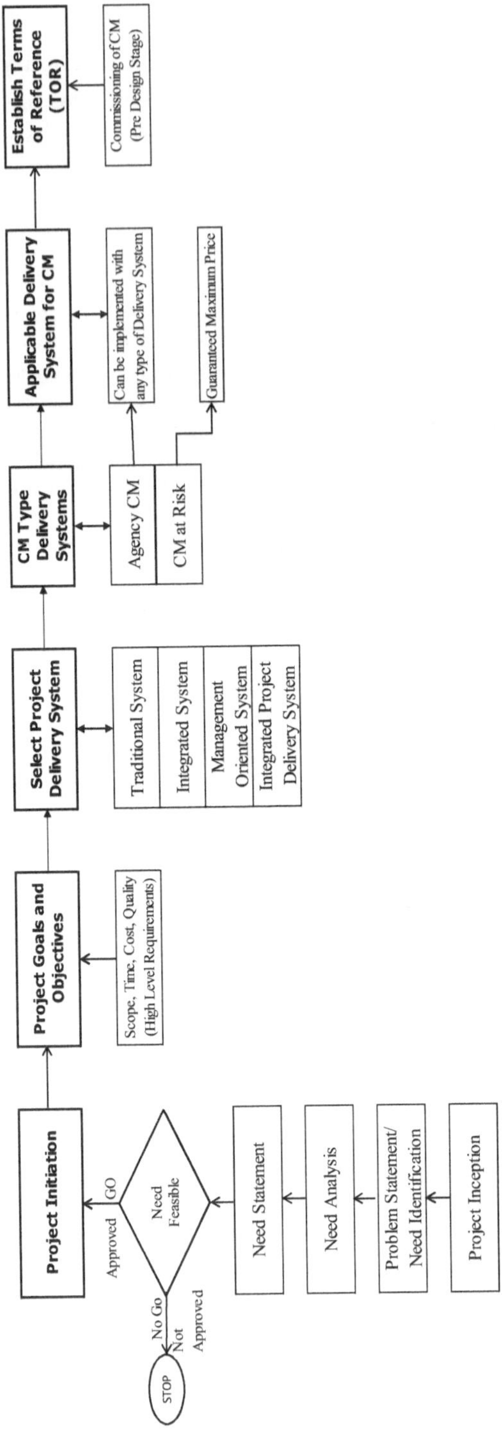

FIGURE 3.1 Roadmap for construction management.

agency. The project development process begins with project initiation and ends with project closeout and finalization of project records after completion of construction of the project. The project initiation starts with the identification of a need and a business case.

The need of the project is created by the owner and is linked to the available financial resources to develop the facility. The owner's needs are quite simple and are based on the following:

- To have the best facility for the money which is to have maximum profit or services at a reasonable cost.
- On-time completion, which is, to meet the owner's/user's schedule.
- Completion within budget, which is, to meet the investment plan for the facility.

The owner's needs must be well defined, indicating the minimum requirements of quality and performance, an approved main budget, and required completion date. Sometimes, the project budget is fixed and therefore the quality of the building system, materials, and finishes of the project need to be balanced with the budget. A business case typically addresses the business need for the project and the value the project brings to the business (project value proposition). A value proposition is a promise of value to be delivered by the project. The following questions address the value proposition:

- How the project solves the current problems or improve the current situation?
- What specific benefits the project will deliver?
- Why the project is the ideal solution for the problem?

Business need assessment is essential to ensure that the owner's business case has been properly considered before the initial project brief (Need Statement) is developed.

TABLE 3.2 lists major points to be considered for the need analysis of a construction project and TABLE 3.3 illustrates the need statement.

Once the owner's need is identified, the traditional approach is pursued through need analysis and then a feasibility study or an economic appraisal of owner needs or benefits, taking into account relevant moral, social, environmental, and technical constraints. The feasibility study takes its starting point from the output of the project identification need. A feasibility study is conducted to assist owner/decision makers in making the decision that will be in the best interest of the owner. Depending on the circumstances, the feasibility study may be short or lengthy, simple, or complex. In any case, it is the principal requirement in project development as it gives the owner an early assessment of the viability of the project and the degree of risk involved. The project feasibility study is usually performed by the owner through his/her own team or by engaging individuals/organizations involved in the preparation of economic and financial studies. However, the feasibility study can be conducted

TABLE 3.2
Need Analysis

Serial Number	Points to be Considered
1	Is the project in line with the organization's strategy/strategic plan and mandated by management in support of a specific objective?
2	Is the project a part of the mission statement of the organization?
3	Is the project a part of the vision statement of the organization?
4	Is the need mandated by regulatory body?
5	Is the need for meeting government regulations?
6	Is the need to fulfill the deficiency/gap of such type of project(s) in the market?
7	Is the need created to meet market demand?
8	Is the need to meet the research and development requirements?
9	Is the need for technical advances?
10	Is the need generated to construct a facility/project which is innovative in nature?
11	Is the need aim to improve the existing facility?
12	Is the need a part of a mandatory investment?
13	Is the need to develop infrastructure?
14	Is the need will serve the community and fulfill social responsibilities?
15	Is the need created to resolve a specific problem?
16	Will the need have an effect on the environment?
17	Does the need have any time frame to implement?
18	Does the need have financial constraints?
19	Does the need have a major risk?
20	Is the need within the capability of the owner/client, either alone or in cooperation with other organizations?
21	Can the need be managed and implemented?
22	Is the need realistic and genuine?
23	Is the need measurable?
24	Is the need beneficial?
25	Does the need comply with environmental protection agency requirements?
26	Does the need comply with government's health and safety regulations?

TABLE 3.3
Need Statement

Serial Number	Points to be Considered
1	Project purpose and need.
	a. Project description
2	What is the purpose of the project?
	a. Project justification
3	Why is the project needed now?

(Continued)

TABLE 3.3 (Continued)
Need Statement

Serial Number	Points to be Considered
4	How was the need of the project determined?
	a. Supporting data
5	Is it important to have the needed project?
6	Whether such facility/project is required?
7	What are the factors contributing to the need?
8	What is the impact of the need?
9	Will the need improve the existing situation and be beneficial?
10	What are the hurdles?
11	What is the timeline for the project?
12	What are the funding sources for the project?
13	What are the benefits of the projects?
14	What are the environmental impacts?

by a specialist consultant in this field. TABLE 3.4 illustrates the qualification of a consultant to perform a feasibility study.

The objective of a feasibility study is to review the technical/financial viability of the project to give sufficient information to enable the client to proceed or abort the project. A feasibility study is undertaken to analyze the ability to complete a project successfully, taking into account various factors such as economic, technological, scheduling, etc. A feasibility study looks into the positive and negative effects of a project before investing the company resources, namely, time and money.

The following are the contents of a feasibility study report:

1. Purpose of feasibility study.
2. Project history (project background information).
3. Description of the proposed project:
   a. Project location.
   b. Plot area.
   c. Interface with adjacent/neighboring areas.
   d. Expected project deliverables.
   e. Key performance indicators.
   f. Constraints.
   g. Assumptions.

4. Business case:
   a. Project need.
   b. Stakeholders.
   c. Project benefits.
   d. Financial benefits.

TABLE 3.4
Qualification of Consultant

Serial Number	Description
1	Experience in conducting a feasibility study.
2	Experience in conducting feasibility studies in similar types and nature of projects.
3	Fair and neutral with no prior opinion about what decision should be made.
4	Experience in strategic and analytical analysis.
5	Knowledge of analytical approach and background.
6	Ability to collect a large number of important and necessary data via work sessions, interviews, surveys, and other methods.
7	Market knowledge.
8	Ability to review and analyze market information.
9	Knowledge of market trends in similar types of projects/facilities.
10	Multidisciplinary experienced team having proven record in the following fields: a. Financial analyst b. Engineering/technical expertise c. Policy experts d. Project scheduling
11	Experience in the review of demographic and economic data.

   e. Estimated cost.
   f. Estimated time.
   g. Justification.

5. Feasibility study details:
   a. Technical.
   b. Economical.
   c. Financial.
   d. Time scale.
   e. Environmental.
   f. Ecological.
   g. Sustainability.
   h. Political, legal.
   i. Social.

6. Risk.
7. Environmental impact(considerations).
8. Social impact (considerations).
9. Final recommendation.

The outcome of the feasibility study helps to select a defined project that meets the stated project objectives, together with a broad plan of implementation. If the feasibility study shows that the objectives of the owner are best met through the ideas generated, then the project is moved to a further stage to deliver the intended objectives of the project passing through different stages of the project life cycle.

3.3.2 Project Goals and Objectives

Project goals and objectives are prepared taking into consideration the final recommendations/outcome of the feasibility study. Clear goals and objectives provide the project team with appropriate boundaries to make decisions about the project and ensure the project/facility will satisfy the owner's/end user's requirements fulfilling the owner's needs. Establishing properly defined goals and objectives is the most fundamental element of project planning. Therefore, the project goals and objectives must be:

- Specific. (Is the goal specific?)
- Measurable. (Is the goal measurable?)
- Agreed upon/achievable. (Is the goal achievable?)
- Realistic. (Is the goal realistic or result-oriented?)
- Time (cost) limited. (Does the goal have a time element?)

3.3.3 Select Project Delivery System

A project delivery system is defined as the organizational arrangement among various participants comprising the owner, designer, contractor, and many other professionals involved in the design and construction of a project/facility to translate/transform the owner's needs/goals/objectives into a finished facility/project to satisfy the owner's/end user's requirements. There are different types of project delivery systems followed in construction projects, however, each of the project delivery systems is a variation of the following basic types:

- Design-Bid-Build (traditional delivery method).
- Design-Build (integrated system).
- Management Oriented System:
  - Project manager.
  - Construction management (CM).
- Integrated Project Delivery System.

3.3.4 Construction Management Delivery System

In the CM type of management process, the owner retains control of design and direction. It has a fully integrated design and construction process thus minimizing changes, disputes, and delays in the completion of the construction project.

There are two forms of construction management systems. These are:

1. Agency Construction Management
2. Construction Management at Risk

3.3.4.1 Agency Construction Management

Agency construction management is a management process where the construction manager acts as an advisor to the owner. The owner may engage the CM for the entire life cycle of the construction project or during a specific phase of a construction project. Normally a CM is engaged as early in the project as possible to guide and assist the owner through all the phases of the project or for a specific phase. The agency construction manager can be used in conjunction with any project delivery system. The agency construction manager is always with the owner. Agency construction management is a fee-based service in which the construction manager is responsible exclusively to the owner and acts in the owner's interest. With an agency CM type of construction management system, the owner holds the contracts directly with the general contractor and also assumes the risks of delivery including cost and schedule, quality, safety, performance, and errors and omissions. The agency CM has no financial stake in the project and does not hold any subcontracts.

The agency CM type of management system can be used with different types of project delivery systems. The following are FIGURES that illustrate the contractual relationship of agency CM with different types of project delivery systems and their corresponding sequential activities.

FIGURE 3.2 illustrates the contractual relationship of the design-bid-build delivery system using Agency CM whereas FIGURE 3.3 illustrates sequential activities for this type of delivery system.

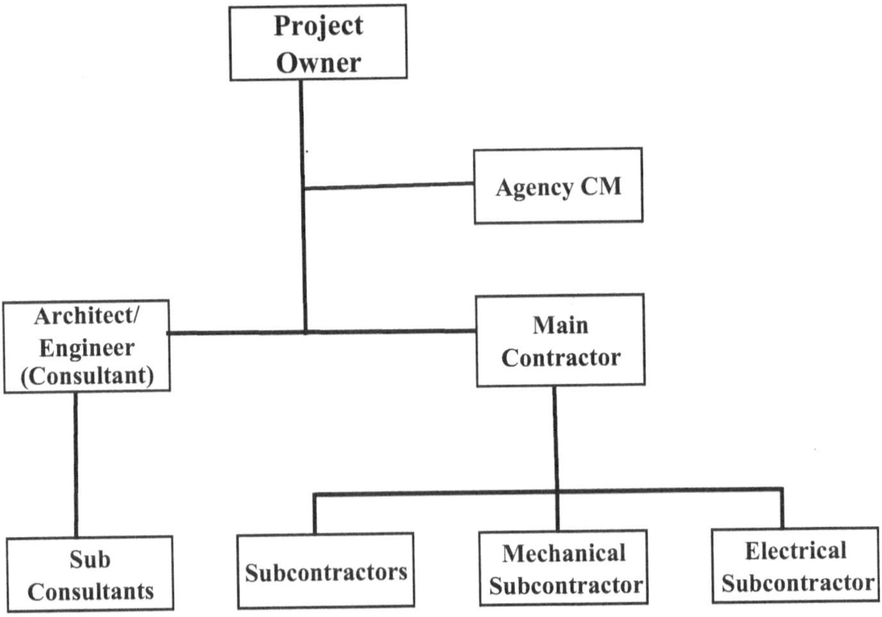

FIGURE 3.2 Design-bid-build with agency CM.

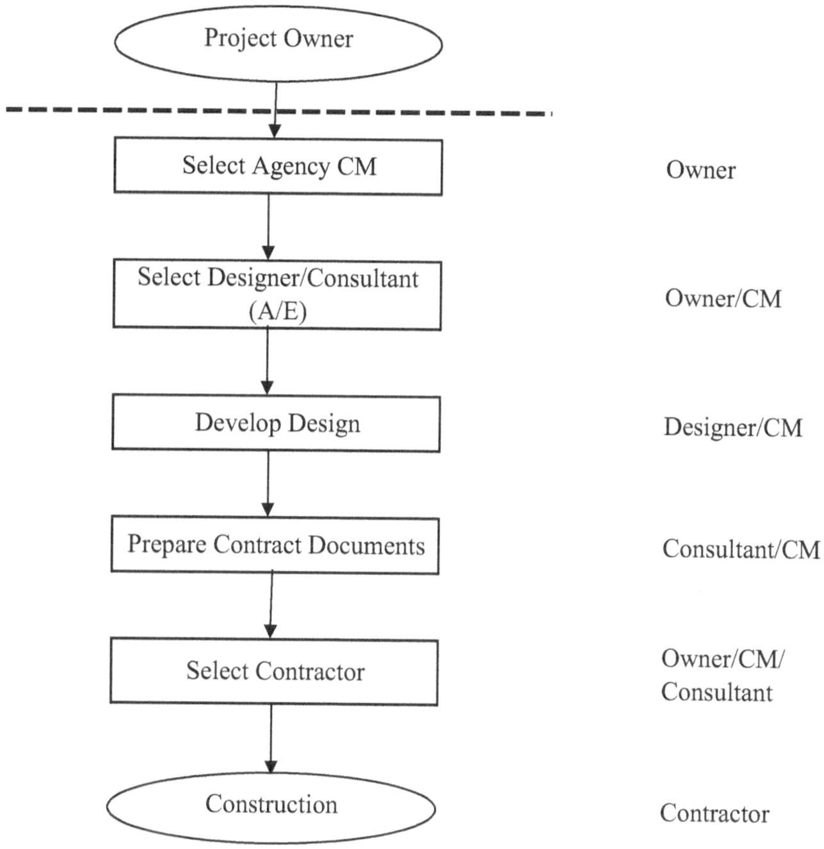

FIGURE 3.3 Sequential activities of agency CM design-bid-build.

FIGURE 3.4 illustrates the contractual relationship of a multiple prime contractor delivery system using agency CM whereas FIGURE 3.5 illustrates sequential activities for this type of delivery system.

FIGURE 3.6 illustrates the contractual relationship of a design-build delivery system using agency CM whereas FIGURE 3.7 illustrates sequential activities for this type of delivery system.

FIGURE 3.8 illustrates the contractual relationship of an agency CM type management system whereas FIGURE 3.9 illustrates sequential activities for this type of delivery system.

FIGURE 3.10 illustrates the contractual relationship of a CM at risk delivery system using agency CM whereas FIGURE 3.11 illustrates sequential activities for this type of delivery system.

3.3.4.2 Construction Management at Risk

CM at risk is a project delivery system. CM at risk is selected based on qualification, experience, and reputation. In this system, the CM enters into a contract with

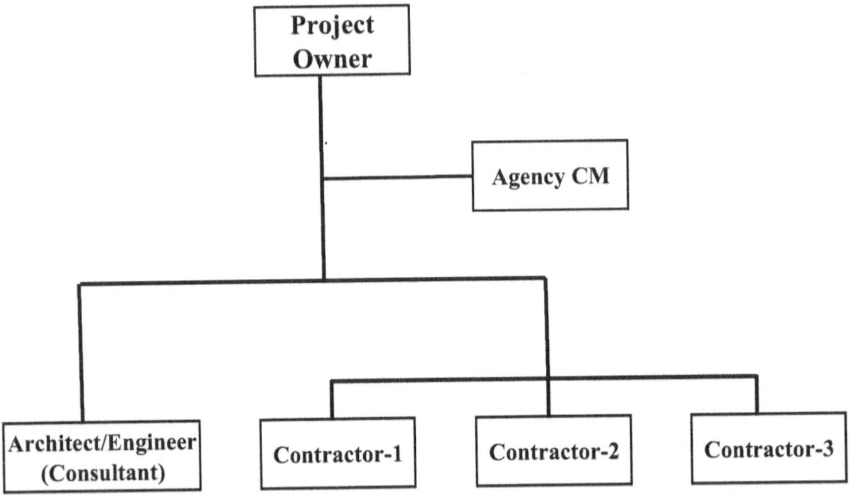

FIGURE 3.4 Multiple prime contractor with agency CM.

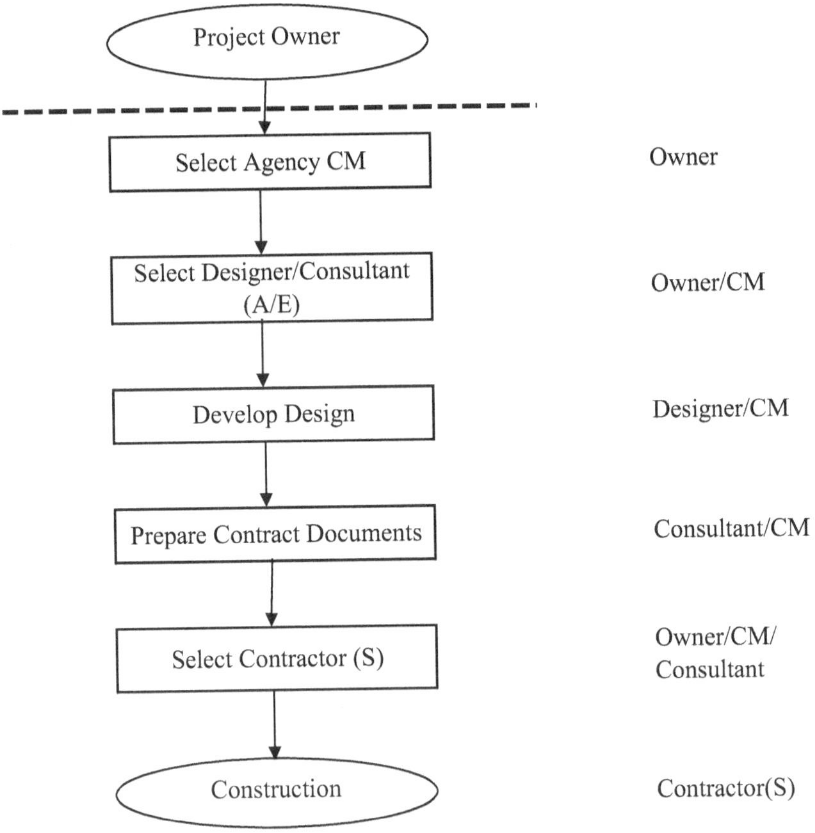

FIGURE 3.5 Sequential activities of agency CM multiple prime contractor.

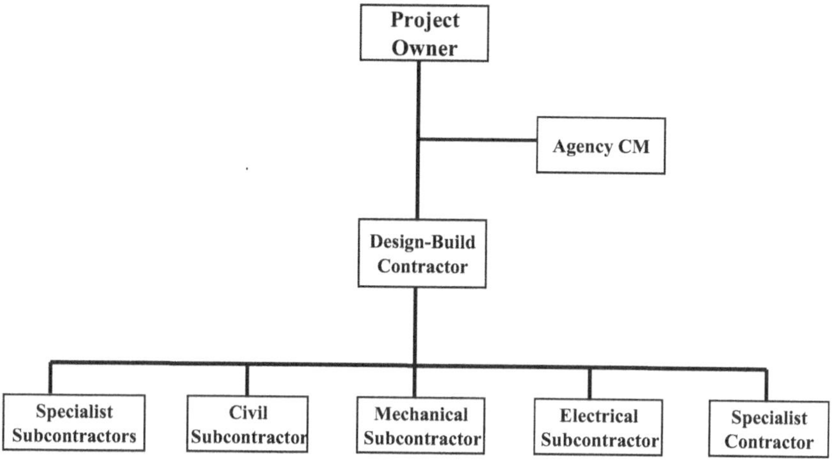

FIGURE 3.6 Design-build with agency CM.

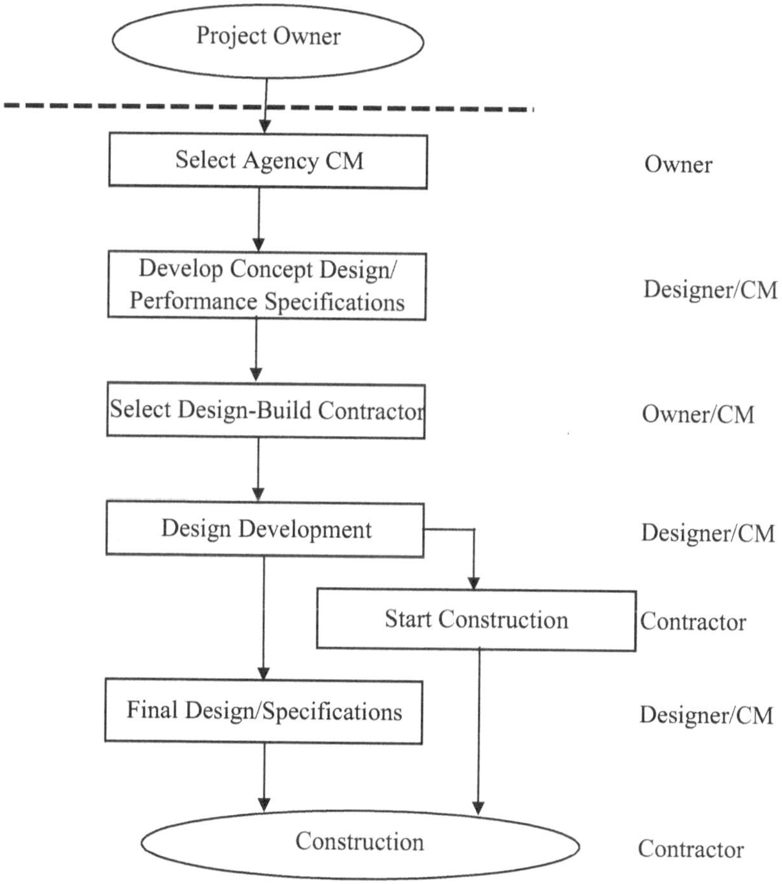

FIGURE 3.7 Sequential activities of agency CM design-build.

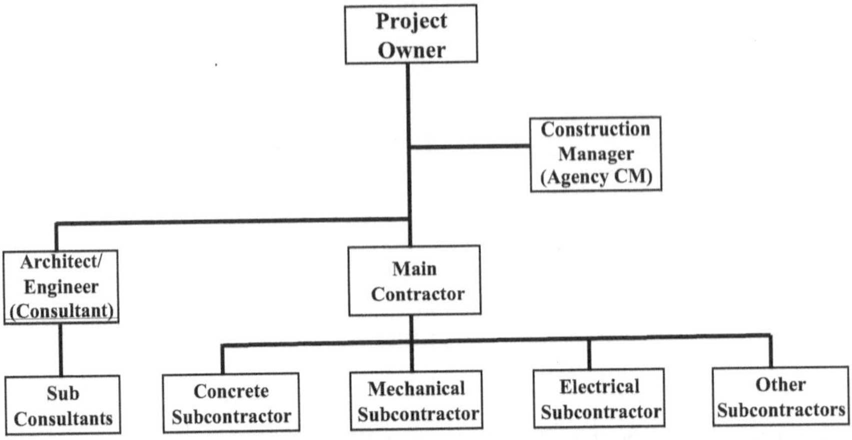

FIGURE 3.8 Contractual relationship of agency CM.

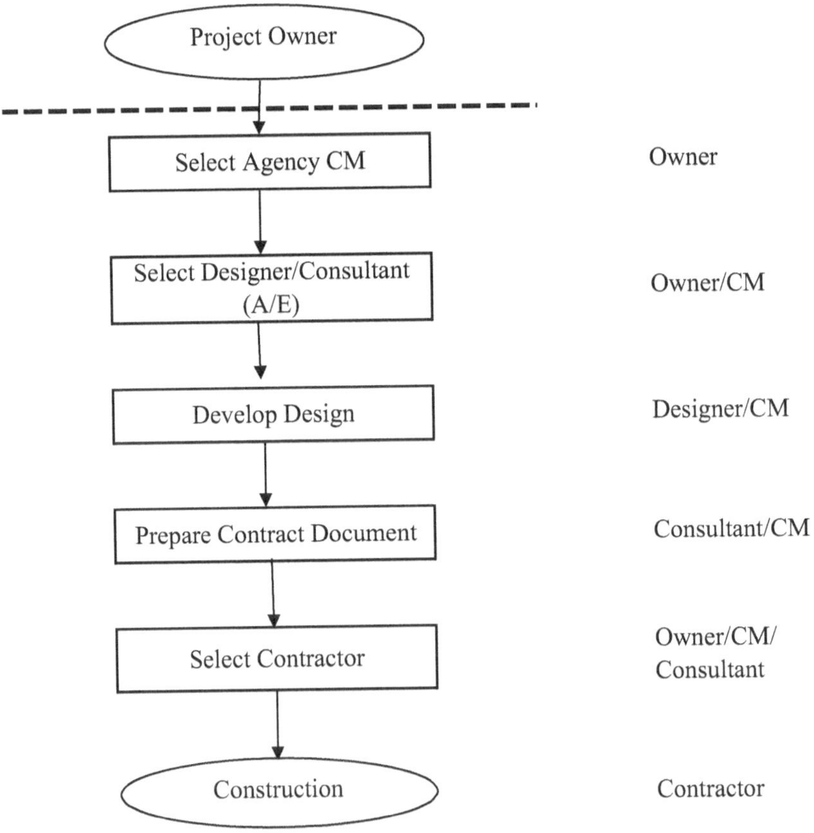

FIGURE 3.9 Sequential activities of agency CM.

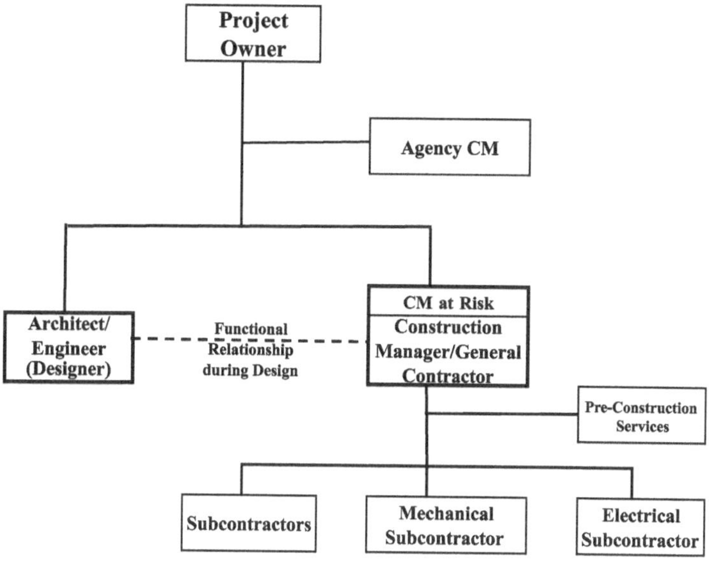

FIGURE 3.10 CM at risk with agency CM.

the owner at an early stage and becomes a member of the collaborative project team with that of the A/E (designer). A CM at risk typically contracts with the owner in two stages. Design services and construction services are contracted separately (versus design-build where the contracts are combined). In the first stage, the CM acts as an advisor to the owner and assists in the development of the conceptual and preliminary design phase. The second phase involves the development of a detailed design and completion of construction for a negotiated fixed OR guaranteed maximum price (GMP). In CM at risk arrangement criteria for the final selection of a contractor is not based on total construction cost (versus design-bid-build delivery system where total construction cost is a factor for final selection). The guaranteed maximum price contracts method is typically used with CM at risk.

In the USA, certain states have promulgated statutory/regulatory guidelines to hire CMs at risk to procure construction services.

FIGURE 3.12 illustrates the contractual relationship of the CM at risk delivery system whereas FIGURE 3.13 illustrates sequential activities for this type of delivery system.

3.3.5 Establish Terms of Reference (TOR)

Normally, Terms of Reference (TOR), also known as a design brief, are prepared by the owner/client or by the project manager on behalf of the owner describing the objectives and requirements to develop the project. In the case of the construction management type of management system, the construction manager assists the owner/client to prepare the TOR.

A client brief (TOR) defines the objectives for the project and agrees on a project brief to guide the next stage of the project. A well-prepared accurate and

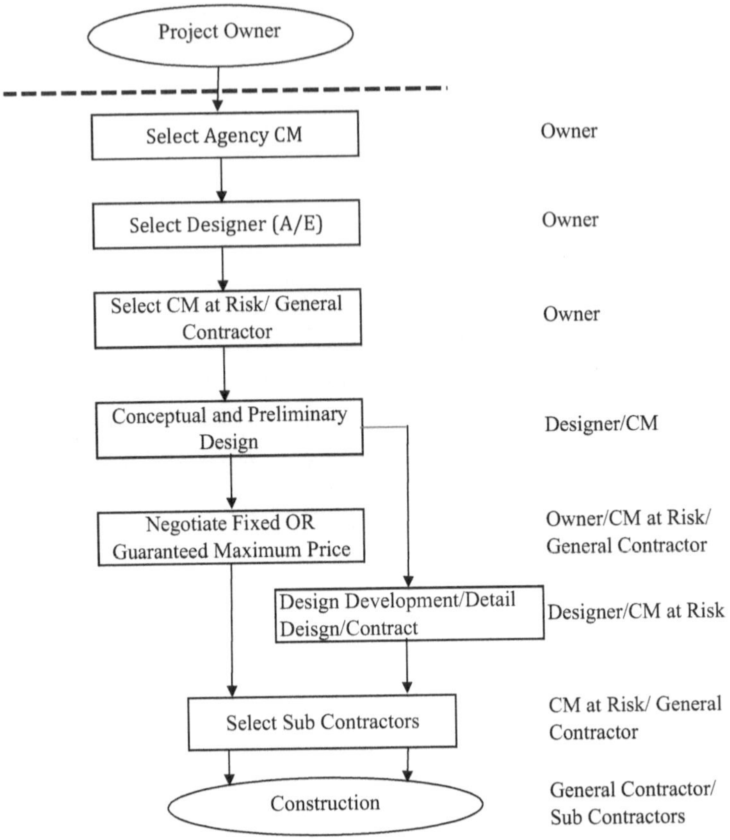

FIGURE 3.11 Sequential activities of agency CM at risk.

comprehensive client brief (TOR) is essential to achieve a qualitative and competitive project. The TOR gives the project team (designer) a clear understanding for the development of the project. Further, the TOR is used throughout the project as a reference to ensure that the established objectives are achieved. Client brief or TOR describes information such as:

- The need or opportunity which has triggered the project.
- Proposed location of the project.
- Project/facility to be developed.
- Project function and size.
- Performance characteristics of the project.
- Procurement strategy.
- Project assumptions and constraints.
- Estimated timescale.
- Estimated cost.
- Initial list of defined risks.
- Description of approval requirements.

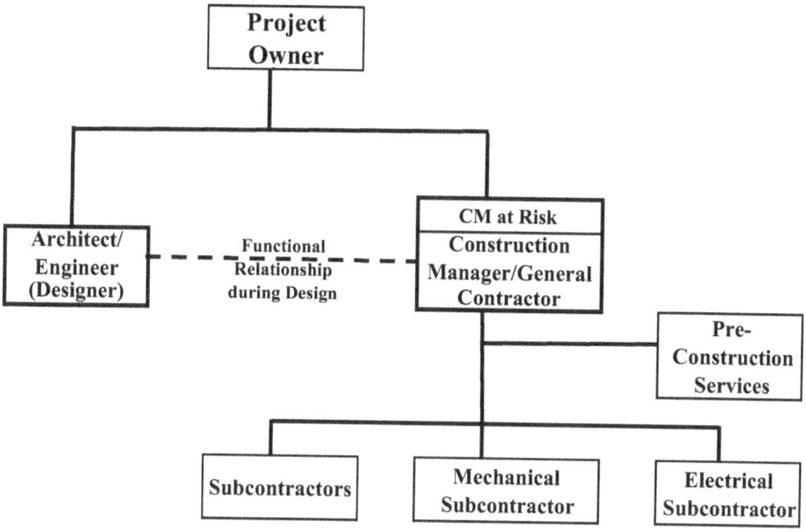

FIGURE 3.12 Construction management at risk contractual relationship.

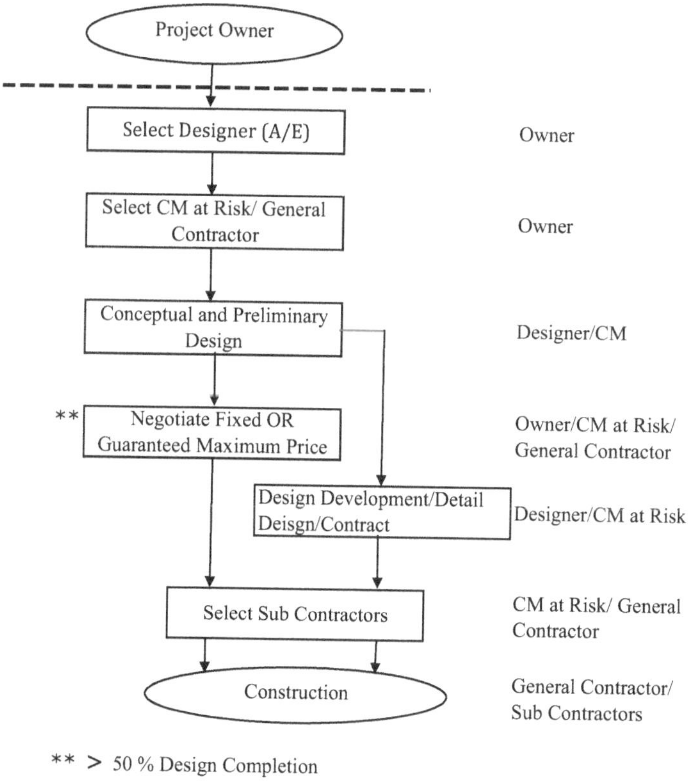

FIGURE 3.13 Sequential activities of CM at risk delivery system.

For the development of construction projects, TOR generally details the services to be performed by the designer (consultant) which include, but are not limited to, the following:

- Pre-development studies, collection of required data, and analysis are the same to prepare design drawings and documents for the project.
- Development of alternatives.
- Preparation of concept design.
- Preparation of schematic design.
- Preparation of detailed design.
- Project deliverables.
- Obtaining authorities approvals.
- Compliance standards, codes, and practices.
- Coordinating and participating in value engineering study.
- Preparation of construction schedule.
- Preparation of construction budget.
- Preparation of contract documents for bidding purposes.
- Prequalification/selection of contractor.
- Evaluation of proposals.
- Recommendation of the contractor to the owner/client.

In the case of a construction management type of system, some of the aforementioned activities are performed by the construction manager as the CM acts as an advisor to the owner.

The following are the requirements for a building construction project, normally mentioned in TOR, to be prepared by the designer during the conceptual phase for submission to the owner:

1. Site Plan:
    a. Civil
    b. Services
    c. Landscaping
    d. Irrigation
2. Architectural Design
3. Building and Engineering Systems:
    a. Structural
    b. Mechanical (HVAC)
    c. Public Health
    d. Fire Suppression Systems
    e. Electrical
    f. Low Voltage Systems
    g. Others
4. Cost Estimates
5. Schedules

3.4 QUALIFICATION OF CONSTRUCTION MANAGER

Construction management is a process wherein professional knowledge, skills, tools, and techniques are applied to project activities to meet the project requirements and successful completion of the project to the satisfaction of the owner/end user. The construction manager should be a highly qualified person having full knowledge of the construction management system to achieve the project objectives. TABLE 3.5 illustrates the typical qualification requirements of a construction manager.

3.5 ROLE OF CONSTRUCTION MANAGER

As per CMAA Owner's Guide to Project Delivery Methods,

> In the past, most owners relied on the experience of the designer to provide a complete and responsible set of contract documents. Recently, more and more owners have found the value in utilizing the advice and expertise of those with overall process, program, and construction management knowledge during the design phase.

TABLE 3.5
Qualifications of Construction Manager

Serial Number	Qualification
1	Thorough knowledge and understanding of construction processes.
2	In-depth knowledge of construction activities.
3	Knowledge of project delivery systems and their adoptability.
4	Knowledge of contracting systems and their suitability for the project.
5	Knowledge of project planning, monitoring, and control.
6	Knowledge of various tools and techniques used in construction projects.
7	Knowledge of construction codes and practices.
8	Knowledge of applicable standards and conditions of contract.
9	Knowledge of information technology.
10	Knowledge and understanding of all disciplines of construction projects.
11	Knowledge of societal needs towards projects.
12	Knowledge and skills to oversee and manage complex construction projects.
13	Knowledge of quality management techniques.
14	Knowledge of HSE practices.
15	Knowledge of risk management.
16	Excellent technical background.
17	Communication skills (oral and written).
18	Strong and responsive leadership skills.
19	Competency in construction management skills.
20	Problem-solving skills.
21	Ability to negotiate.
22	Innovative and creative thinking.
23	Collaborative thinking.
24	Able to motivate and encourage subordinates and other team members to work as a group.
25	Possess high-level university degree educational qualification.
26	Member of related professional institution/society.
27	Basic knowledge of design practices.

Whether provided through owner staffing or a third-party firm, the CM should be engaged as early in the project as possible to guide and assist the owner through all phases of delivering the project. The CM may also act as the owner's representative with the other members of the project team, being the point of contact for the designer, contractor, and any other specialty consultants engaged in the project by the owner.

The construction management delivery system is a project procurement/contract management process whereby the construction manager (firm or an individual with sound project manager skills) undertakes to manage the carrying out of the work through contractors be they general or trade specific and oversee the performance throughout the project life cycle by systematic application of set of management skills and principles.

The role of a construction manager is to apply comprehensive management and control effort to the project beginning at the early project planning stages and continuing until project completion. The construction management process involves the application and integration of comprehensive project controls to the design and construction process to achieve successful project delivery/completion.

The roles and responsibilities of a construction manager in a project may vary substantially and can be performed under a variety of contractual terms.

Regardless of the project delivery system utilized in the construction projects, the construction manager can play a pivotal role in all the phases of the construction project life cycle.

Generally, construction projects are divided into three stages, which are:

1. Preconstruction:
   a. Study (predesign)
   b. Design
   c. Bidding

2. Construction
3. Post-Construction

The owner may engage the construction manager to oversee all the activities during all the phases of the construction project life cycle or engage the CM to perform a specific role during a specific phase of the project depending on the owner's desires and requirements for delegating responsibilities and authority. CM is paid a fee to act as the owner's agent or advisor for a construction project. The following are the roles played by the CM during various stages of a construction project.

3.5.1 CM Role During Pre-Design Stage

During this stage, the CM acts in an advisory role and acts purely in the Agency CM relationship.

During this stage the CM provides the following services:

- Defining overall performance requirements.
- Defining overall project program.

- Develop project scope of work (terms of reference).
- Development of project procedures and standards.
- Establishing a management information system.
- Development of project schedule.
- Development of project budget.
- Identify critical constituents of the project.
- Identify authorities' approvals and permits required.
- Selection of project delivery method.
- Selection of designer (consultant).

3.5.2 CM Role During Design Stage

During the design stage, the CM works collaboratively as well as an independent member of the design team. The CM has to provide information, both data and recommendations that will enable the owner to make the best design decision possible as the owner is responsible for the contractor(s) for design errors (Spearin Doctrine).

The CM as an agent to the owner provides the following services during the design stage:

- Oversee and coordinate design.
- Recommend alternative solutions.
- Life cycle cost analysis.
- Review design documents.
- Constructability review.
- Make suggestions to improve constructability.
- Construction-related recommendations.
- Coordination with regulatory authorities to obtain permits and licenses.
- Recommend selection of materials, building systems, and equipment.
- Value engineering suggestions.
- Value enhancement suggestions.
- Construction schedule.
- Construction cost estimates.
- Budget and schedule management.
- Project risk management.
- Help the designer (A/E) develop multiple bid packages to expedite construction process.
- Review of final drawings and specifications.
- Preparing and revising project management plan.

3.5.2.1 CM Role During Bidding

The role of CM during the bidding stage:

- Preparation of bid packages.
- Prequalification of bidders.
- Establish bidding schedule.
- Management of bidding documents.

- Pre-Bid meetings to familiarize bidders with the bidding documents.
- Answering all queries related to bid.
- Addendum to bid documents.
- Receive bids.
- Bid review and evaluation.
- Bid analysis and comparison of bids.
- Contract negotiation.
- Prepare recommendation to owner to accept or reject the proposals.
- Participate in contractor selection.
- Conduct pre-award meeting(s) with the selected contractor.
- Incorporate addenda changes into contract documents.
- Prepare construction contract.
- Notice to proceed.

3.5.3 CM Role During Construction Stage

The construction manager's roles during the construction stage vary depending on the type of contract the owner has entered with the CM. There are two forms of construction management processes.

1. Agency CM
2. CM at risk

3.5.3.1 Agency CM

As agency CM the construction manager is responsible to perform services related to the following areas of activities:

- Ensure that contractor submitted performance bond.
- Ensure that contractor submitted worker's insurance policy.
- Selecting and recommending contractor's core staff.
- Selecting and recommending subcontractor.
- Establishing and implementing procedures for processing and approval of shop drawings.
- Material approvals.
- Managing contractor's request for information (RFI).
- Change order management.
- Construction schedule approval.
- Construction supervision.
- Quality management.
- Coordination of on-site, off-site inspections.
- Inspection of works.
- Construction contract administration.
- Conduction periodic progress meetings.
- Preparation of minutes of meeting and distribution as per agreed upon matrix.
- Document control.

- Technical correspondence between contractor.
- Managing submittals.
- Monitor daily progress.
- Monitoring contractor's performance and ensuring that the work is performed as specified, as per approved shop drawings, and as per applicable codes.
- Scope control.
- Construction scheduling and monitoring.
- Cost tracking and management.
- Review, evaluation, and documentation of claims.
- Maintain project progress record.
- Evaluation of payment requests and recommending progress payments.
- Monitor project risk.
- Monitoring contractor's HSE plans (health, safety, and environment).
- Coordination of work by multiple contractors.
- Coordination for delivery and storage of owner-supplied materials, systems.
- Testing of systems.
- Punch list.

3.5.3.2 CM at Risk

Under the CM at risk form of project delivery system, the CM acts as a general contractor. The owner transfers the responsibility and risk to the CM at risk contractor for the entire construction effort, "performance risk" including subcontract administration and coordination.

3.5.4 CM Role During Testing, Commissioning, and Handover Stage

- Testing and commissioning of systems.
- Review of as-built drawings.
- Review of record documents and manuals.
- Warranties and guarantees.
- Authorities' approval for occupancy.
- Coordination of handover-takeover procedure.
- Move in plan.
- Punch list.
- Preparing list of lessons learned.
- Substantial completion.
- Archiving project documents.
- Settlement of claims.
- Project final account.
- Project close-out.

4 Construction Management Tools

Abdul Razzak Rumane

4.1 INTRODUCTION

Quality tools are charts, check sheets, diagrams, graphs, techniques, and methods that are used to create an idea, engender planning, analyze the cause, analyze the process, foster evaluation and create a wide variety of situations for continuous quality improvement. Applications of quality tools enhance chances of success and help maintain consistency, accuracy, increase efficiency, and process improvement.

4.2 CATEGORIZATION OF TOOLS

There are several types of tools, techniques, and methods, in practice, which are used as quality improvement tools and have a variety of applications in the manufacturing and process industry. However, all of these tools are not used in construction projects due to the nature of construction projects which are customized and non-repetitive. Some of these quality management tools that are most commonly used in the construction industry are listed:

1. Classic quality tools.
2. Management and planning tools.
3. Process analysis tools.
4. Process improvement tools.
5. Innovation and creative tools.
6. Lean tools.
7. Cost of quality.
8. Quality function deployment.
9. Six Sigma.
10. TRIZ.

A brief description of these tools is given in Sections 4.2.1 through 4.2.10.

4.2.1 Classic Quality Tools

Classic quality tools have a long history. These tools are listed in TABLE 4.1.

DOI: 10.1201/9781003636687-4

TABLE 4.1
Classic Tools of Quality

Sr. No.	Name of Quality Tool	Usage
Tool 1	Cause and effect diagram	To identify possible cause and effect in processes.
Tool 2	Check sheet	To provide a record of quality. How often it occurs?
Tool 3	Control chart	A device in statistical process control to determine whether or not the process is stable.
Tool 4	Flow chart	Used for graphical representation of a process in sequential order.
Tool 5	Histogram	Graphs used to display frequency of various ranges of values of a quantity.
Tool 6	Pareto chart	Used to identify the most significant cause or problem.
Tool 7	Pie chart	Used to show classes or groups of data in proportion to the whole set of data.
Tool 8	Run chart	Used to show measurement against time in a graphical manner with a reference line to show the average of the data.
Tool 9	Scatter diagram	To determine whether there is a correlation between two factors.
Tool 10	Stratification	Use to show the pattern of data collected from different sources.

All of these tools have been in use since World War II. Some of these tools date prior to 1920. The approach includes both quantitative and qualitative aspects, which taken together, focus on company-wide quality.

A brief definition of these quality tools is as follows (values shown in the figures are indicative only):

4.2.1.1 Cause and Effect Diagram

The cause and effect diagram is also called an Ishikawa diagram, after its developer Kaoru Ishikawa, or fishbone diagram. It is used to identify possible causes and effects in the process. It is used to explore all the potential or real causes that result in a single output. The causes are organized and displayed in a graphical manner to their level of importance or detail. It is a graphical display of multiple causes with a particular effect. The causes are organized and arranged mainly into four categories. These are:

1. Machine
2. Manpower
3. Material
4. Method

The effect or problem being investigated is shown at the end of a horizontal arrow. Potential causes are shown as labeled arrows entering the main cause arrow. Each arrow may have a number of other arrows entering as the principal cause or factors are reduced to their sub-causes. FIGURE 4.1 illustrates an example cause and effect diagram for the rejection of executed (installed) false ceiling work by the supervision engineer for not complying with contract specifications.

4.2.1.2 Check Sheet

A check sheet is a structured list, prepared from the collected data to indicate how often each item occurs. It is an organized way of collecting and structuring data. The purpose of a check sheet is to collect the facts in the most efficient manner. Data is collected and ordered by adding tally or check marks against predetermined categories of items or measurements. FIGURE 4.2 illustrates a check sheet for checklist approval records for wire bundles.

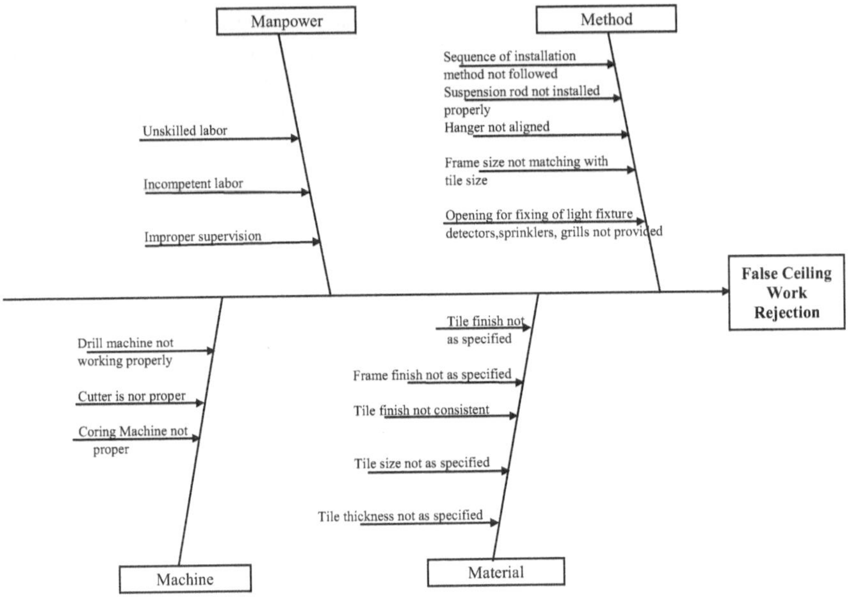

FIGURE 4.1 Cause and effect diagram.

Approval Record for Wire Bundles

	Approved	Not Approved	Total	% Not Approved
1.5 mm² Wire	𝍸𝍸 //	///	50	4
2.5 mm² Wire	𝍸𝍸𝍸 𝍸 //	///	85	6
4.0 mm² Wire	𝍸𝍸𝍸 𝍸 //	///	25	3
6.0 mm² Wire	𝍸 //	///	15	2
10.0 mm² Wire	𝍸 //	///	10	2

FIGURE 4.2 Check sheet.

4.2.1.3 Control Chart

The control chart is the fundamental tool of statistical process control. The control chart is a graph used to analyze how a process behaves over time and to show whether the process is stable or is being affected by special causes of variation and creating an out-of-control condition. It is used to determine whether the process is stable or varies between the predictable limits. It can be employed to distinguish between the existence of a stable pattern of variation and the occurrence of an unstable pattern. With control charts, it is easy to see both special and common cause variations in a process. There are many types of control charts. Each is designed for a specific kind of process or data. FIGURE 4.3 illustrates a control chart for the distribution of an air handling unit (AHU).

4.2.1.4 Flowchart

The flowchart is a pictorial tool used for the representation process in sequential order. The flowchart uses graphic symbols to depict the nature and flow of the steps in a process. It helps to see whether the steps of a process are logical, uncover the problems, or miscommunications, define the boundaries of a process, and develop a common base of knowledge about a process. The flow of steps is indicated with arrows connecting the symbols. Flow charts can be applied at all stages of the project life cycle. FIGURE 4.4 illustrates a flowchart for contractor's staff approval in construction projects.

4.2.1.5 Histogram

A histogram is a pictorial representation of the frequency distribution of data. It is created by grouping the data points into a cell and displays how frequently different values occur in the data set. FIGURE 4.5 illustrates a histogram for employee reporting time.

4.2.1.6 Pareto Chart

Pareto chart is named after Vilfredo Pareto, a nineteenth-century Italian economist who postulated that a large share (80%) of wealth is owned by a small (20%)

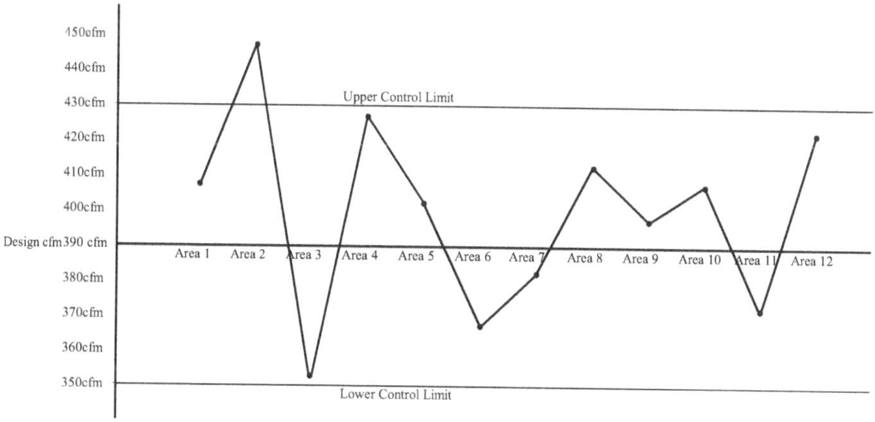

FIGURE 4.3 Control chart.

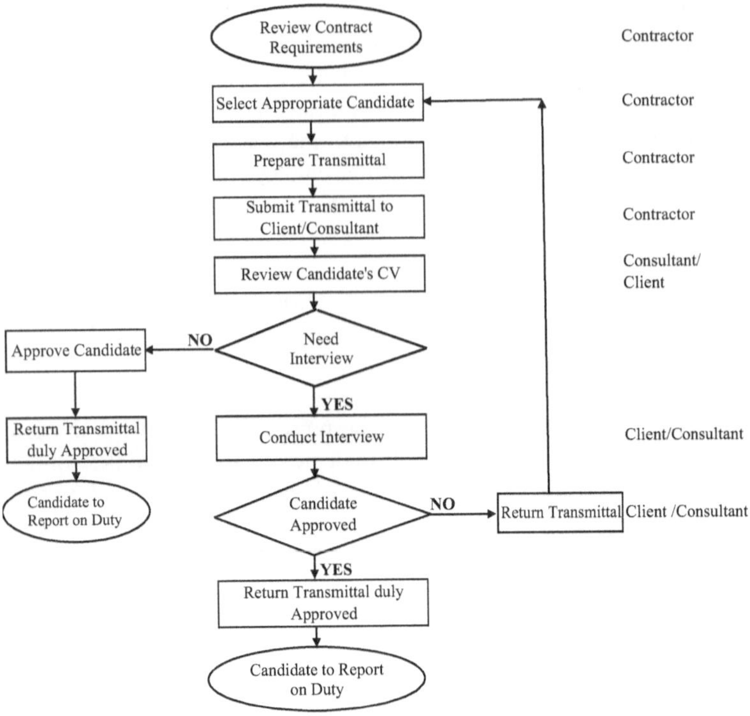

FIGURE 4.4 Flowchart.

Employee Reporting Time Histogram

TIME	No. of Employee
730-745	5
745-800	125
800-815	65
815-830	45
830-900	2

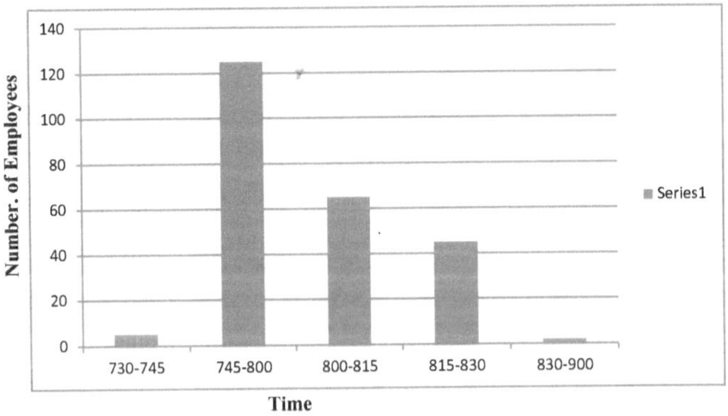

FIGURE 4.5 Histogram.

percentage of the population. A Pareto chart is a graph bar having a series of bars whose height reflects the frequency of occurrence. Pareto charts are used to display the Pareto Principle in action and arrange data so that the few vital factors that are causing most of the problems reveal themselves. The bars are arranged in descending order of height from left to right. Pareto charts are used to identify those factors that have the greatest cumulative effect on the system, and thus less significant factors can be screened out from the process. A Pareto chart can be used at various stages in a quality improvement program to determine which step to take next. FIGURE 4.6 illustrates the Pareto chart for division cost of a construction project.

4.2.1.7 Pie Chart

A pie chart is a circle divided into wedges to depict the proportion of data or information in order to understand how they make up the whole. The entire pie chart represents all the data, while each slice or wedge represents a different class or group within the whole. The portions of the entire circle or pie sum to 100%. FIGURE 4.7 illustrates the contents of the contractor's site staff at the construction project site.

4.2.1.8 Run Chart

The run chart is a graph plotted by showing measurement (data) against time. Run charts are used to know the trends or changes in a process variation over time over the average and also to determine if the pattern can be attributed to common causes of variation, or if special causes of variation were present. A run chart is also used to monitor process performance. Run charts can be used to track improvements that have been put in place, checking to determine their success. FIGURE 4.8 illustrates

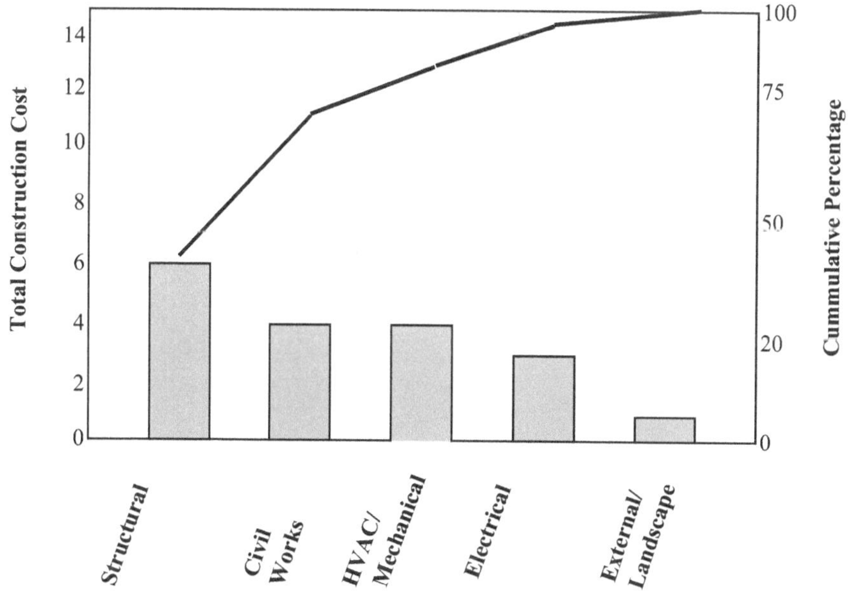

FIGURE 4.6 Pareto chart.

Site Staff

Managers	2
Engineers	16
Technicians	9
Formen	20
Skilled	230
Unskilled	570

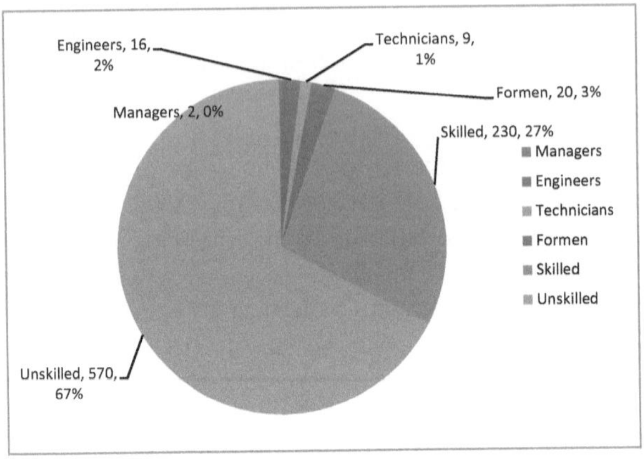

FIGURE 4.7 Pie chart.

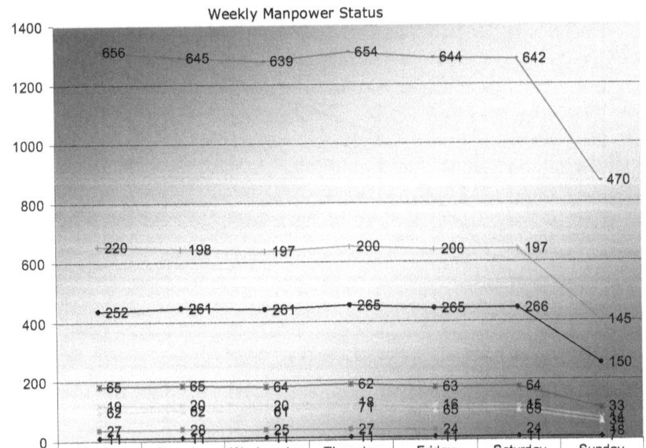

	Monday	Tuesday	Wednesday	Thursday	Friday	Saturday	Sunday
Total	656	645	639	654	644	642	470
Others	220	198	197	200	200	197	145
Mechanical &Electrical	252	261	261	265	265	266	150
Marble	65	65	64	62	63	64	33
Steel Structure & Aluminum	19	20	20	18	16	15	14
False Ceiling & Partition	62	62	61	71	65	65	34
Plaster & Paint	27	28	25	27	24	24	18
Steel fixer &Carpenter	11	11	11	11	11	11	7

FIGURE 4.8 Run chart.

a run chart for weekly manpower of different trades of a project. It is similar to control chart but does not show control limits

4.2.1.9 Scatter Diagram

A scatter diagram is a plot of one variable versus another. It is used to investigate the possible relationship between two variables that both relate to the same event. It helps to know how one variable changes with respect to another. It can be used to identify potential root cause of problems and to evaluate cause-and-effect relationships. FIGURE 4.9 illustrates a scatter diagram for beam quantity of various lengths.

4.2.1.10 Stratification

Stratification is a graphical representation of data collected from different data. FIGURE 4.10 illustrates the stratification diagram for cable drums.

4.2.2 Management and Planning Tools

Seven management tools are the most popular after-quality classic tools. These tools are listed in TABLE 4.2.

These tools are focused on managing and planning quality improvement activities.

These tools are focused on managing and planning quality improvement activities. A brief definition of these quality tools is as follows (values shown in the figures are indicative only):

4.2.2.1 Activity Network Diagram

An activity network diagram (AND) is a graphical representation chart showing the interrelationship among activities (tasks) associated with a project. The activity

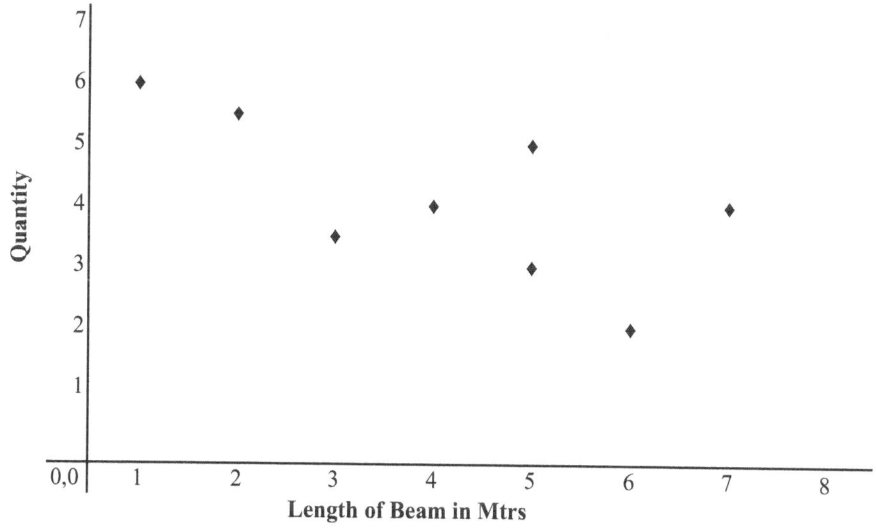

FIGURE 4.9 Scatter diagram.

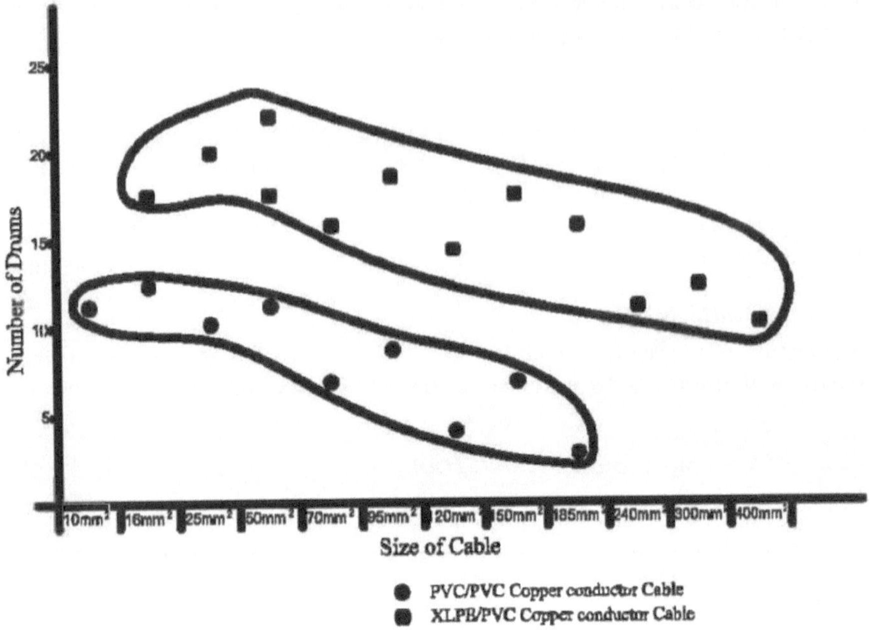

FIGURE 4.10 Stratification.

TABLE 4.2
Management and Planning Tools

Sr. No.	Name of Quality Tool	Usage
Tool 1	Activity network diagram (arrow diagram)/critical path method	Used when scheduling or monitoring tasks is complex or lengthy and has schedule constraints.
Tool 2	Affinity diagram	Used to organize a large group of items in smaller categories that are easier to understand and deal with.
Tool 3	Interrelationship diagram (relations diagram)	Used to show logical relationships between ideas, process, cause, and effect.
Tool 4	Matrix diagram	Used to analyze the correlations between two or more groups of information.
Tool 5	Prioritization matrix	Used to choose one or two options that have important criteria from several options.
Tool 6	Process decision program chart	Used to help contingency plan.
Tool 7	Tree diagram	Used to break down or stratify ideas in progressively more detailed steps.

network diagram was developed by the U.S. Department of Defense. It was first used as a management tool for military projects. It was adapted as an educational tool for business managers. In the activity network diagram, each activity is represented by one and only one arrow in the network and is associated with an estimated time to perform the activity. The activity network diagram analyzes the sequences of tasks necessary to complete the project. The direction of the arrow specifies the order in which the events must occur. The event represents a point in time that indicates the completion of one or more activities and the beginning of new ones. FIGURE 4.11 illustrates the activity network diagram.

There are two kinds of network diagrams, the "Activity on Arrow" (AoA) network diagram and the "Activity on Node" (AoN) network diagram.

Arrow diagrams or activity on arrows (AoA) use a diagramming method to represent the activities on arrows and connect them at nodes (rectangles) to show the dependencies. With the AoA method, the detailed information about each activity is placed on arrow or as footnotes at the bottom.

FIGURE 4.12 illustrates the arrow diagramming method for concrete foundation work.

Activities originating from a certain event cannot start until the activities terminating the same event have been completed. This is known as precedence relations. These relationships are drawn using the precedence diagramming method (PDM). The PDM technique is also referred to as "Activity on Node" (AoN) because it shows the activities in a node (box) with arrows showing dependencies. "Activity on Node" network diagrams have the activity information written in small boxes that are the nodes of the diagram. Arrows connect the boxes to show the logical relationships between pairs of activities.

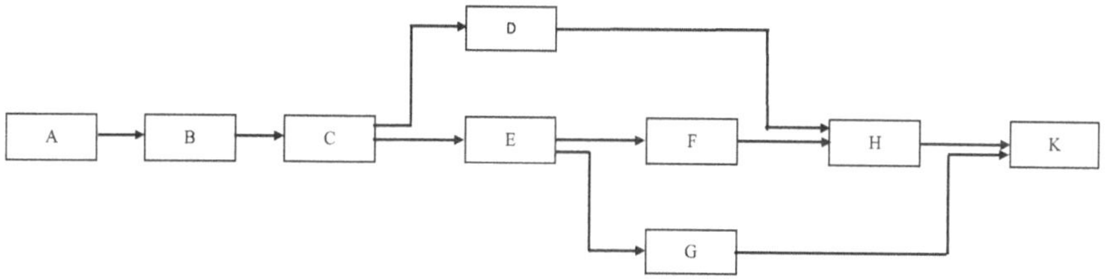

FIGURE 4.11 Activity network diagram.

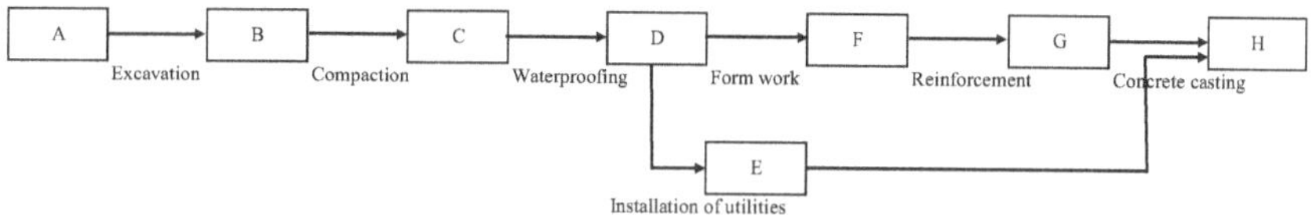

FIGURE 4.12 Arrow diagram.

In the networking diagram, all activities are related in some direct way and may be further constrained by indirect relationships. The following are direct logical relationships or dependencies among project-related activities:

1. Finish to Start—Activity A must finish before activity B can begin.
2. Start to Start—Activity A must begin before activity B can begin.
3. Start to Finish—Activity A must begin before activity B can finish.
4. Finish to Finish—Activity A must finish before activity B can finish.

Apart from these, there are other dependencies such as:

1. Mandatory
2. Discretionary
3. External

FIGURE 4.13 illustrates dependency relationship diagrams and FIGURE 4.14 illustrates the precedence diagramming method.

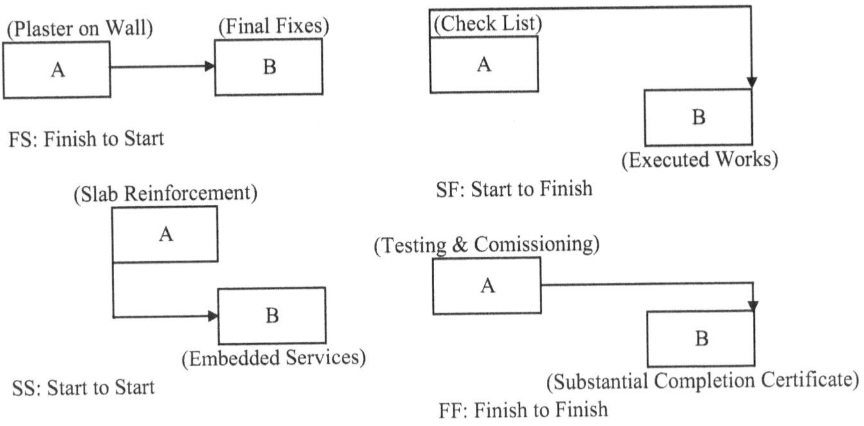

FIGURE 4.13 Dependency relationship.

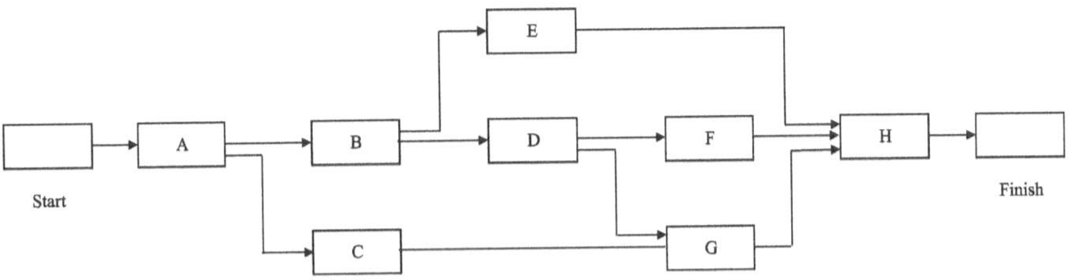

FIGURE 4.14 Precedence diagramming method.

The Activity Network Diagram (AND) or arrow diagram is a tool used for detailed planning, a tool for analyzing schedules during execution, and a tool for controlling complex or large-scale projects. A network diagram uses nodes and arrows. Date information is added to each activity node.

The Critical Path Method (CPM) chart is an expanded activity network diagram showing an estimated time to complete each activity and connecting these activities based on the task to be performed. The critical path is a sequence of interrelated predecessor/successor activities that determines the minimum completion time for a project. The duration of the critical path is the sum of the activities' duration along the path. The activities in the critical path have the least scheduling flexibility. Any delays along the critical path would imply that additional time would be required to complete the project. There may be more than one critical path among all the project activities, so completion of the entire project could be affected due to delaying activity along any of the critical paths. TABLE 4.3 illustrates the activity relationship for the substation project and FIGURE 4.15 critical path method (CPM) diagram for the construction of the substation.

TABLE 4.3
Activity Relationship for Substation

Activity Number	Description of Activity	Duration in Days	Preceding Activity
1	Start	0	
2	Mobilization	21	1
3	Preparation of Site	15	2
4	Staff Approval	15	2
5	Material Approval	15	2,4
6	Shop Drawing Approval	15	2,5
7	Procurement (Structural work)	15	5
8	Procurement (Pipes, Ducts, Sleeves)	7	5
9	Procurement (HT Switchgear, Transformer)	60	5
10	Procurement (Civil, MEP, Furnishing)	30	5
11	Excavation	4	3,6
12	Blinding Concrete	4	5,7,11
13	Raft Foundation	7	5,6,7,12
14	Utility Services (Embedded)	4	8,12
15	Concrete (Floor)	1	6,13,14
16	Trenches	7	13
17	Embedded Services/Ducts in Trench	2	8,13
18	Concrete (Transformer Area)	1	16,17
19	Walls and Columns	7	15,18
20	Formwork for Slab	3	6,7,19
21	Reinforcement	2	7,20
22	Embedded Services	1	8,20

(Continued)

TABLE 4.3 (Continued)
Activity Relationship for Substation

Activity Number	Description of Activity	Duration in Days	Preceding Activity
23	Concrete (Roof Slab)	1	21,22
24	Masonry Work	14	6,10,23
25	Installation of Equipment	7	9,18,24
26	Installation of Electromechanical Items	14	6,10,23,24
27	Installation of Ventilation System	4	10,23,24
28	Finishes	7	10,24
29	Installation of Final Fixes	3	10,28
30	Furnishing	2	10,28
31	Testing of Equipment	2	25
32	Testing of HVAC, Firefighting, Electrical System	4	25,26,27
33	Handing Over	1	29,30,31,32
34	End	0	

Note: The duration is indicative only to understand sequencing and the relationship of activities.

The activity network diagram is also known as the Program Evaluation and Review Technique (PERT). PERT is used to schedule, organize, and coordinate tasks within a project.

PERT planning involves the following steps:

1. Identify specific activity.
2. Identify milestones of each activity.
3. Determine the proper sequence of each activity.
4. Construct a network diagram.
5. Estimate the time required to complete each activity. The three-point estimation method using the following formula can be used to determine the approximated estimate time for each activity:

Estimated time = (Optimistic + 4*Most likely + pessimistic)/6

6. Compute the Early Start (ES), Early Finish (EF), Late Start (LS), and Late Finish (LF) times for each activity in the network.
7. Determine the critical path.
8. Identify the critical path for possible schedule compression.
9. Evaluate the diagram for milestones and target dates in the overall project.

FIGURE 4.16 illustrates the Gantt chart.

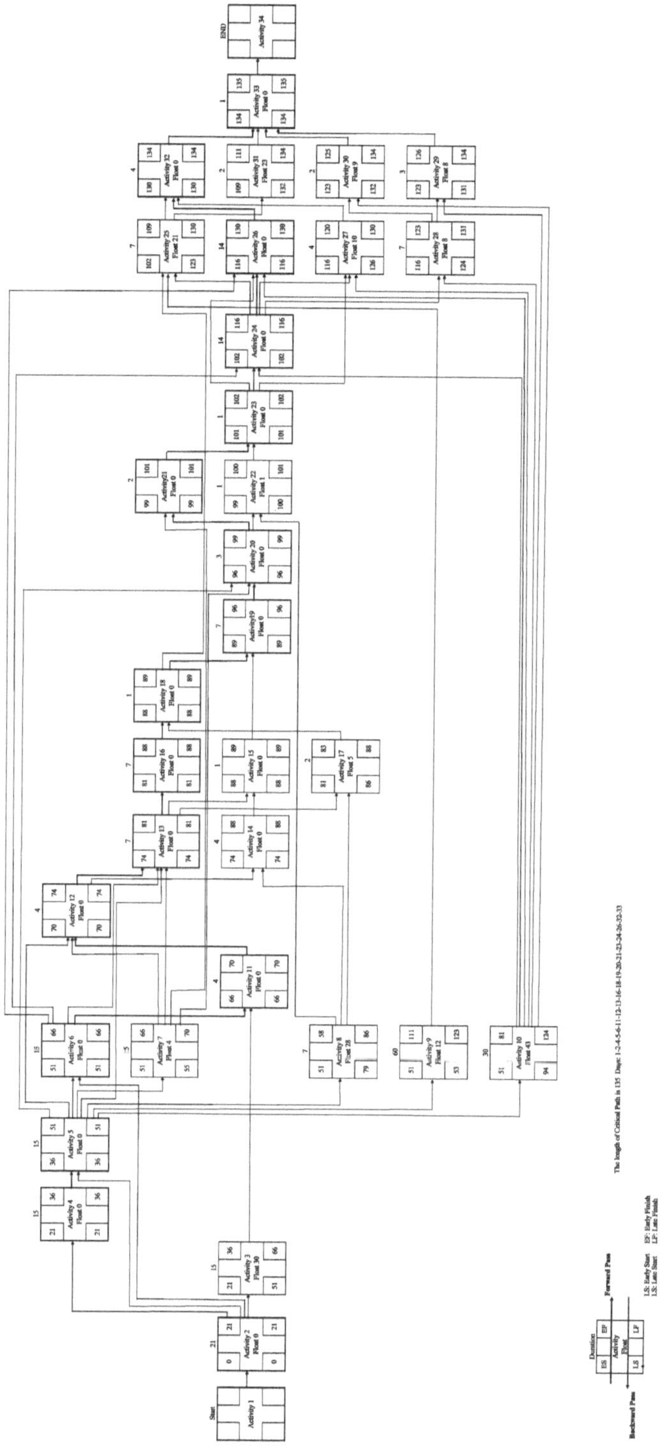

FIGURE 4.15 Critical path method.

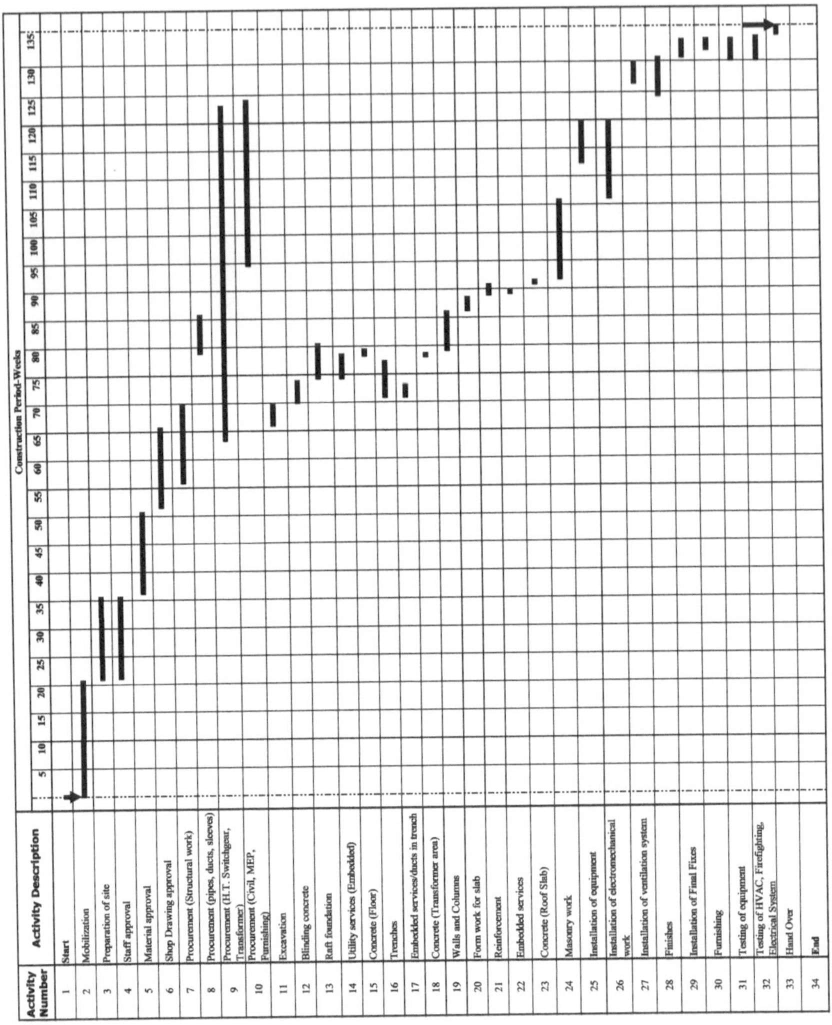

FIGURE 4.16 Gantt chart.

4.2.2.2 Affinity Diagram

An affinity diagram is a tool that gathers a large group of ideas/items and organizes them into a smaller grouping based on their natural relationships. The affinity diagram is a refinement of brainstorming ideas into smaller groups which can be dealt with more easily and satisfy the team members. The affinity process is often used to group ideas generated by brainstorming. The affinity diagram is created as per the following steps:

1. Generate ideas and list the ideas without criticism.
2. Display the ideas in a random manner.
3. Sort the ideas and place them into multiple groups.
4. Continue until smaller groups satisfy all the members.
5. Draw an affinity diagram.

FIGURE 4.17 illustrates the affinity diagram for a concrete slab.

4.2.2.3 Interrelationship Digraph

Interrelationship digraph (di is for directional) is an analysis tool that allows the team members to identify logical causes and effect relationships between the ideas. It is drawn to show all the different relationships between factors, areas, or processes. The digraph makes it easy to pick out the factors in a situation that are driving many of the other symptoms or factors. While affinity diagrams organize and arrange the ideas into groups, interrelationship digraphs identify problems to define the ways in which ideas influence one another. Interrelationship digraph is used to identify cause and effect relationships with the help of directional arrows among critical issues. The number of arrows coming into the node determines the outcome (key indicator) while the outgoing arrows determine the cause (driver) of the issue.

FIGURE 4.18 illustrates the interrelationship digraph for causes of bridge collapse.

4.2.2.4 Matrix Diagram

A matrix diagram is constructed to analyze systematically the correlations between two or more groups of items or ideas. The matrix diagram can be shaped in the following ways:

1. L-Shaped
2. T-Shaped
3. X-Shaped
4. C-Shaped
5. Inverted Y-Shaped
6. Roof-Shaped

Each shape has its own purpose.

1. L-Shaped is used to show interrelationships between two groups or processes.
2. T-Shaped is used to show the relation between three groups. For example, consider there are three groups A, B, and C. In a T-Shaped matrix group,

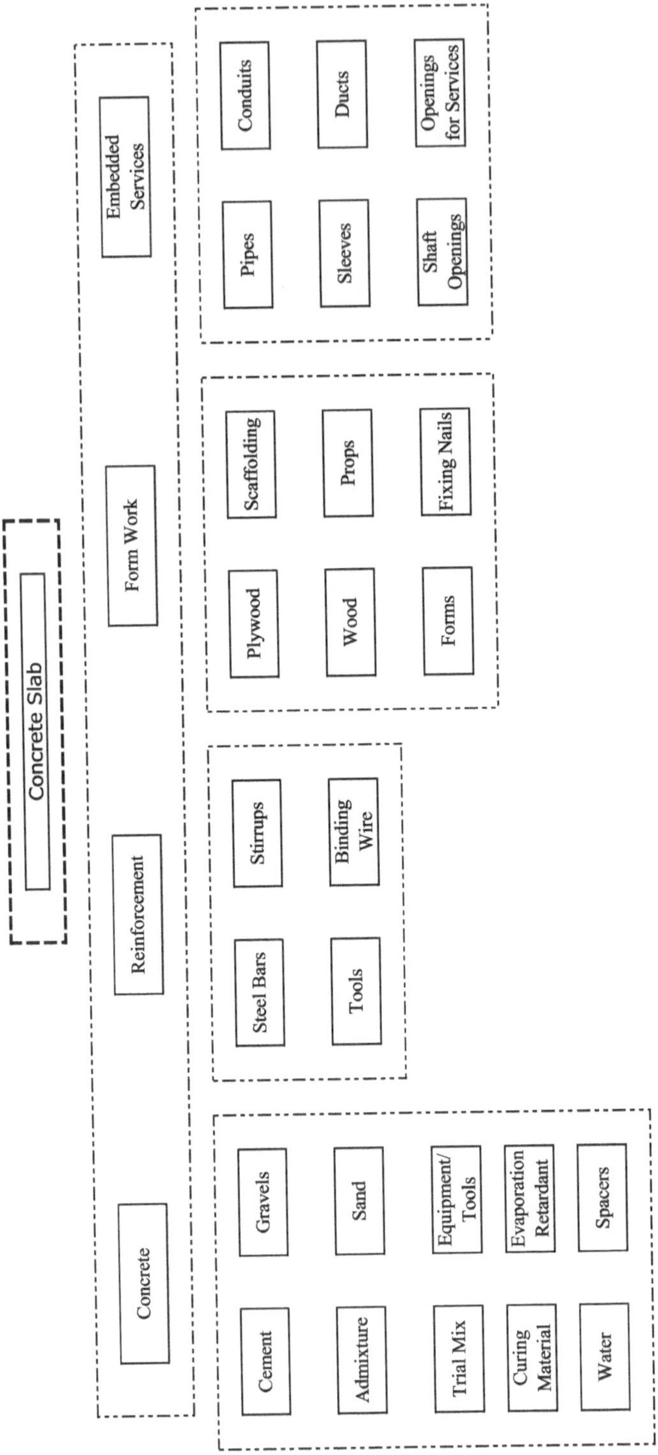

FIGURE 4.17 Affinity diagram.

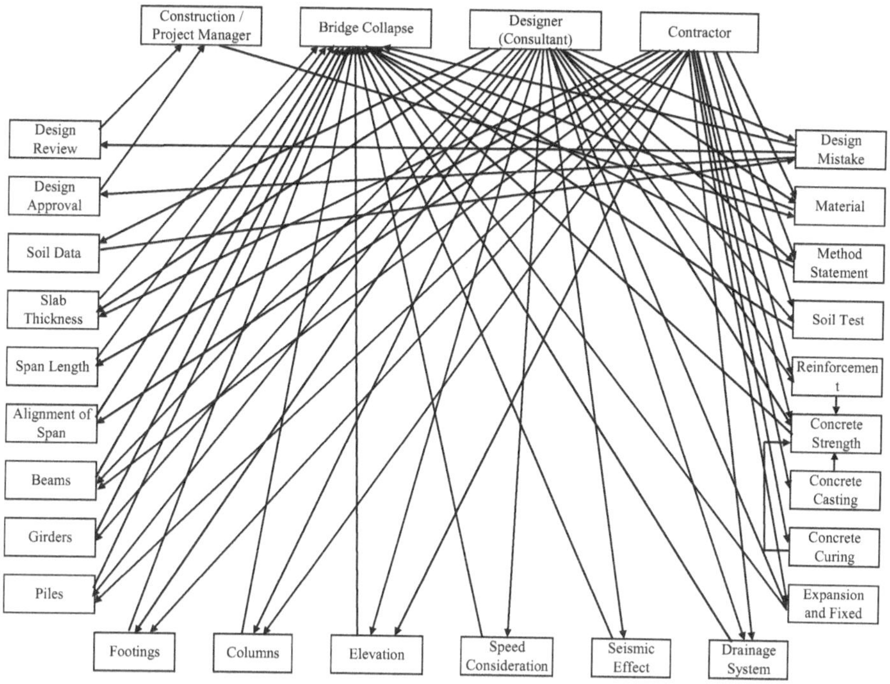

FIGURE 4.18 Interrelationship digraph.

A and B are each related to group C whereas groups A and B are not related to each other.

3. X-Shaped matrix is used to show the relationship among four groups. Each group is related to two other groups in a circular fashion.
4. C-Shaped matrix interrelates three groups of processes or ideas in three-dimensional ways.
5. Inverted Y-Shape matrix is used to show the relation between three groups. Each group is related to other groups in a circular fashion
6. A roof-shaped matrix relates one group of items to itself. Roof-Shaped matrix is used with L- or T-Shaped matrix.

TABLE 4.4 illustrates an L-Shaped matrix.
FIGURE 4.19 illustrates a T-Shaped matrix.

4.2.2.5 Prioritization Matrix

The prioritization matrix assists in choosing between several options in order of importance and priority. It helps decision-makers determine the order of importance considering the relative merit of each of the activities or goals being considered. The prioritization matrix focuses the attention of team members to those key issues and options that are more important for the organization or project.

FIGURE 4.21 illustrates the prioritization matrix.

TABLE 4.4
L-Shaped Matrix

Serial Number	Customer Requirements of Distribution Boards: Component Details	Customer A	Customer B	Customer C
1	Isolator	1	1	-
2	Molded Case Circuit Breaker (MCCB)	-		1
3	HRC Fuse	1	1	-
4	Earth Leakage Circuit Breaker	2	1	2
5	Miniature Circuit Breaker (MCB)	18	12	18
6	Single Bus Bar	-	1	-
7	Double Bus Bar	2	-	2
8	Enclosure with Lock	Yes	No	Yes
9	Surface Mounting	Yes	No	Yes
10	Flush Mounting	-	Yes	

Manufacturing Plant	Products				
Customer					
International				#	#
European Manufacturer	≠	#	#		
Local Plant	#	≠	≠		
# Large Capacity ≠ Small Capacity	600 KVA Transformer	1000 KVA Transformer	1250 KVA Transformer	1600KVA Transformer	2000KVA Transformer
ABC Company	#	≠	≠		
XYZ Company	≠	#	#		
Others				#	#

FIGURE 4.19 T-Shaped matrix.

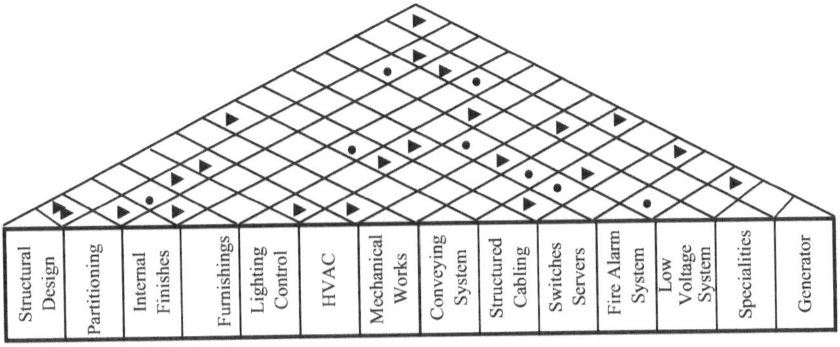

FIGURE 4.20 Roof-Shaped matrix.

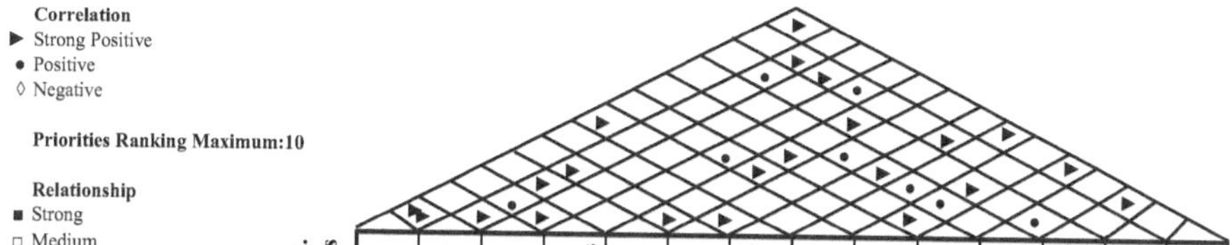

Customer Needs	Tech. Specs Priorities	Structural Design	Partitioning	Internal Finishes	Furnishings	Lighting Control	HVAC	Mechanical Works	Conveying System	Structured Cabling	Switches Servers	Fire Alarm System	Low Voltage System	Specialities	Generator
No. of Class Rooms	10	■	■		■				■	■			■		
Laboratory	9	■	■	□			□			■				■	
Library	7	■		□		□				□			□	■	
Workshops	8	■		■	■		□	■						■	
Auditorium	7	■		■	■	□	□			□			□	■	

FIGURE 4.21 Prioritization matrix.

4.2.2.6 Process Decision Program

The process decision program is a technique used to help prepare contingency plans. The process decision program systematically identifies what might go wrong in a project plan or project schedule and describes specific actions to be taken to prevent the problems from occurring in the first place and to mitigate or avoid the impact of the problems if they occur.

FIGURE 4.22 illustrates the process decision program for submission of contract documents.

4.2.2.7 Tree Diagram

Tree diagrams are used to break down or stratify ideas progressively into more detailed steps. A tree diagram breaks broader ideas into specific details and helps make decisions easier to select the alternative. It is used to figure out all the various tasks that must be undertaken to achieve a given objective.

FIGURE 4.23 illustrates a tree diagram for water in the storage tank.

4.2.3 Process Analysis Tools

TABLE 4.5 illustrates process analysis tools.

4.2.3.1 Benchmarking

Benchmarking is the process of measuring the actual performance of the organization's products, processes, and services and to compare to the best-known industry standards to assist the organization in improving the performance of their products, processes, and services. Benchmarking involves analyzing an existing situation, identifying, and measuring factors critical to the success of the product or services, comparing them with other businesses, analyzing the results, and implementing an action plan to achieve better performance. The following is the process for Benchmarking.

1. Collect internal and external data on work, process, method, product characteristics, and system selected for Benchmarking.

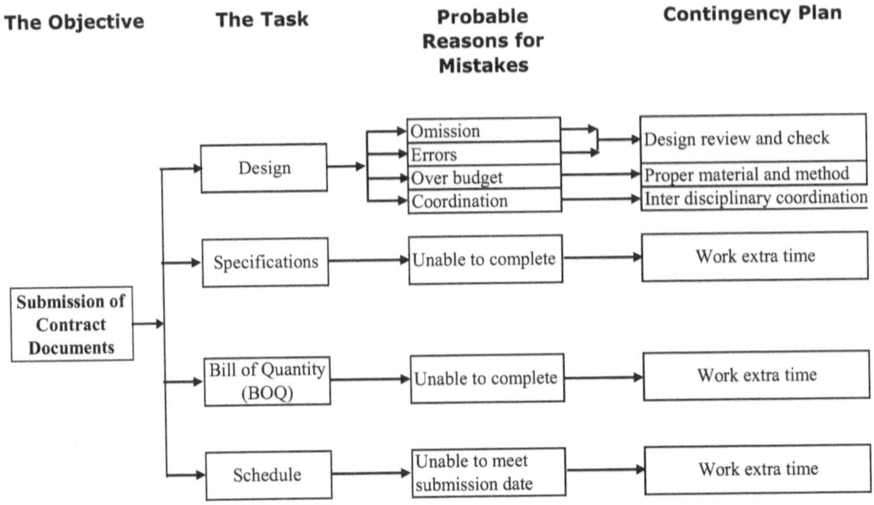

FIGURE 4.22 Process decision diagram.

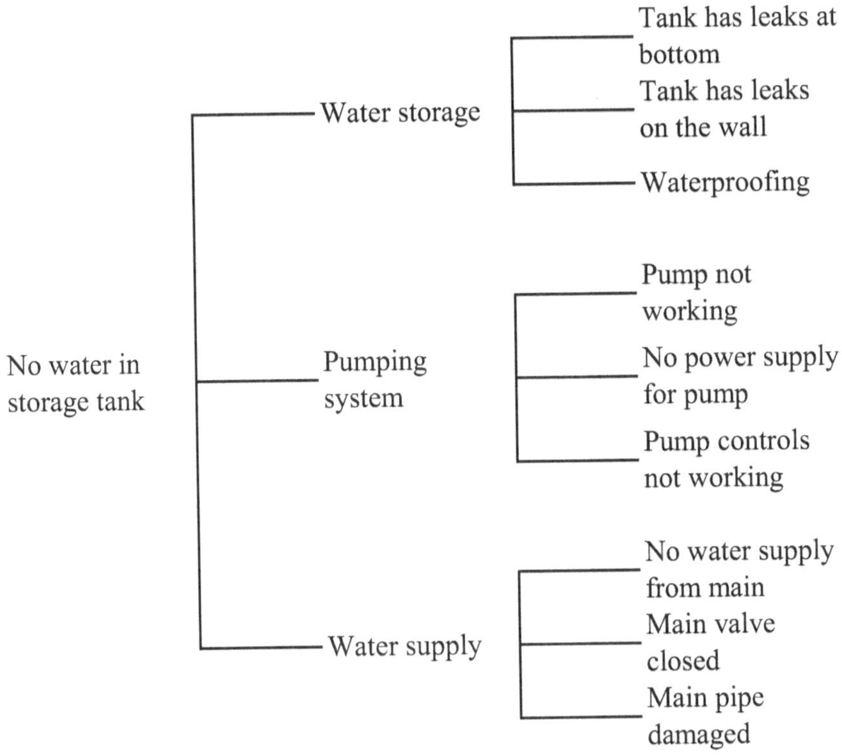

FIGURE 4.23 Tree diagram.

TABLE 4.5
Process Analysis Tools

Sr. No.	Name of Quality Tool	Usage
Tool 1	Brainstorming	Used to generate multiple ideas.
Tool 2	Delphi Technique	Used to get ideas from a select group of experts.
Tool 3	5W2H	The questions used to understand why things happen the way they do.
Tool 4	Mind Mapping	Used to create a visual representation of many issues that can help get more understanding of the situation.
Tool 5	Nominal Group Technique	Used to enhance brainstorming by ranking the most useful ideas.
Tool 6	Six Sigma DMADV	Used primarily for the invention and innovation of modified or new products, services, or processes.
Tool 7	TRIZ	Used to provide systematic methods and tools for analysis and innovative problem-solving.

2. Analyze data to identify performance gaps and determine causes and differences.
3. Prepare an action plan to improve the process in order to meet or exceed the best practices in the industry.
4. Search for the best practices among market leaders, competitors, and non-competitors that lead to their superior performance.
5. Improve performance by implementing these practices.

FIGURE 4.24 illustrates the benchmarking process.

4.2.3.2 Cause and Effect Diagram

It is one of the quality classic tools. It is used to analyze the cause and effect of defects or nonconformance and effect on the process due to these causes.

FIGURE 4.25 illustrates the cause and effect diagram for rejection of masonry work.

4.2.3.3 Cost of Quality

The cost of quality is discussed in detail under Section 4.2.7 of this chapter.

4.2.3.4 Critical to Quality

Critical to quality is a significant step in the design process of a product or service to identify the customer's/client's expectation to fulfill their needs and requirements.

4.2.3.5 Failure Mode and Effects Analysis (FMEA)

Failure mode and effects analysis is to identify all the possible failures in the design of a product, process, or service and their effects on the product, process, and service. Its aim is to reduce the risk of failure and improve the process.

FIGURE 4.26 illustrates failure mode and effects analysis process and FIGURE 4.27 illustrates an example form used to record FMEA readings.

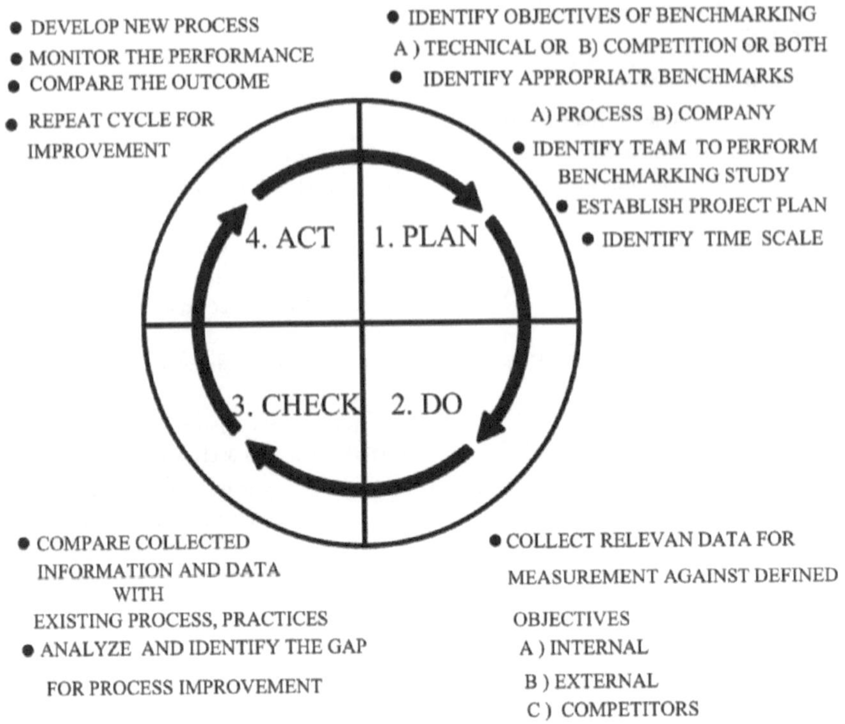

FIGURE 4.24 Benchmarking.

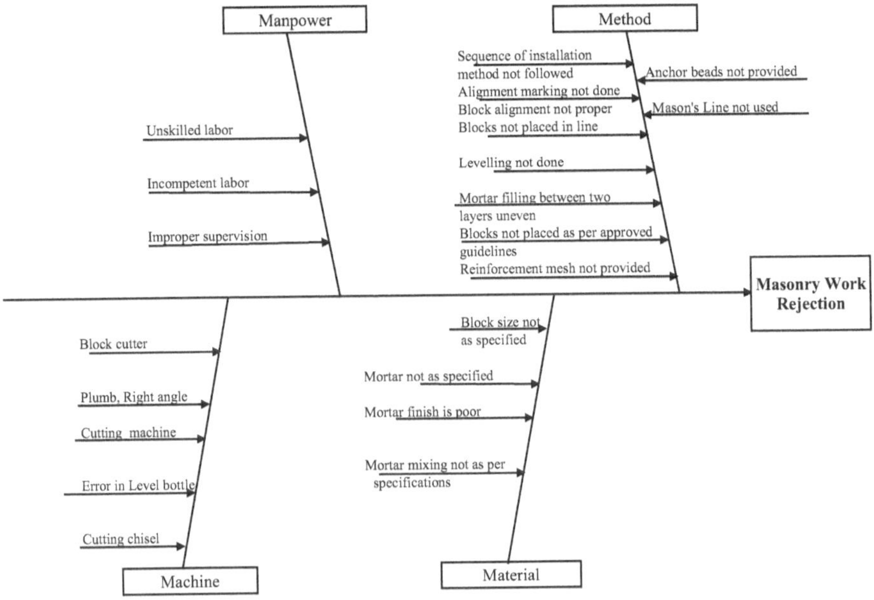

FIGURE 4.25 Cause and effect diagram.

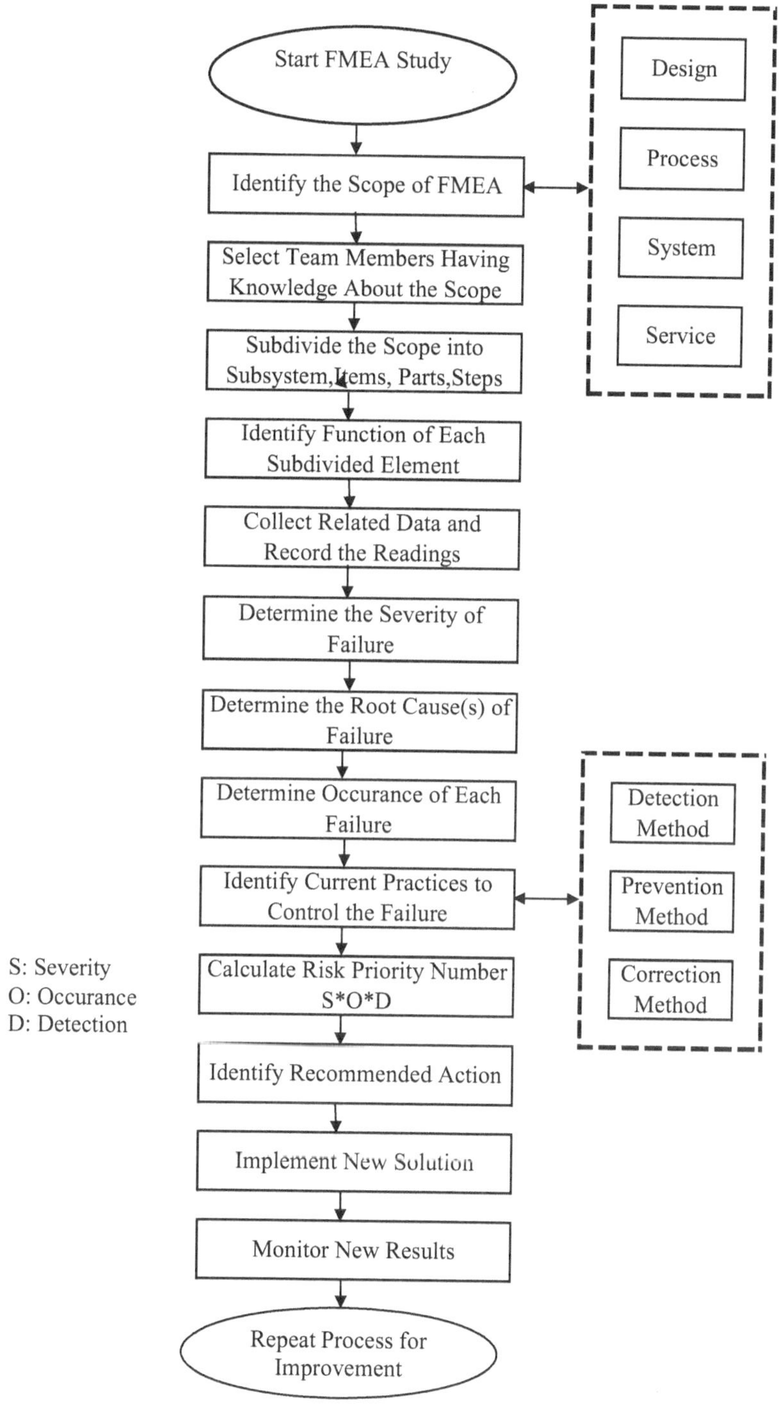

FIGURE 4.26 Failure mode and effects analysis process.

Failure Mode and Effect Analysis

Product Name:

Drawing Reference:
Revision:

EXAMPLE ANALYSIS

Team Members
1
2
3
4

Operation Number	Scope Description	Subdivided Elements	Failure Mode	Effects of Failure	Severity Rating	Cause of Failure	Occurance Rating	Current Practice of Controls			Detection Rating	Current Status of Product				Recommended Corrective Action	Action By	Action Taken Date	Revise Status			
								Detection	Prevention	Correction		S	O	D	RPN				S	O	D	RPN
1	Emergency generator system	Operation	Generator failed to start	1. No lights 2.Life support equipment stopped functioning 3. No power supply fot IT System 4. Water supply pumps stopped 5. No power supply for lift 6. Fire mode operation equipment not operate 5. HVAC system stopped	7 10 9 6 4	1. No signal from ATS 2. Automatic stsrting system failed 3. Low Battery voltage for Starter motor 4. Circuit breaker in off position 5. No diesel in day tank		1.Through BMS 2. No regular check 3. Manual 4. Manual	1. Not in practice	1. Manual						1. Regular check of starting system 2. Check starter regularly 3.Check diesel level regularly. Interface level indicator with BMS 4. Check breaker position regularly. Interface with BMS	Mintenance Engineer					

Legend:-
RPN: Risk Priority Number
S: Severity
O: Occurance
D: Dtetection

ATS: Automatic Transfer Switch
BMS: Building Management System

FIGURE 4.27 FMEA form.

4.2.3.6 5 Whys Analysis

It is used to analyze and solve any problem where the root cause is unknown.

TABLE 4.6 illustrates the five whys analysis chart for the burning of a cable.

4.2.3.7 5W2H

5W2H is about asking questions to understand a process or problem.

The five Ws are:

1. Why?
2. What?
3. When?
4. Where?
5. Who?

And the two Hs are:

1. How many?
2. How much?

TABLE 4.7 illustrates 5W2H for slab collapse.

TABLE 4.6
Why Analysis

Serial Number	Why	Related Analyzing Question
1	Why	Why did the cable burn?
2	Why	Why did the earth leakage relay not trip?
3	Why	Why did the circuit breaker not trip?
4	Why	Why was the poor insulation of the cable not noticed?
5	Why	Why was under-size rating of breaker with respect to current-carrying capacity not noticed?

TABLE 4.7
5W2H

Serial Number	Why	Related Analyzing Question
1	Why	Why did the slab collapse?
2	What	What is the reason for the collapse?
3	Who	Who is responsible?
4	Where	Where is the mistake?
5	When	When did the slab collapse?
6	How many	How many persons affected (injured or died)?
7	How much	How much loss in terms of cost and time?

4.2.3.8 Process Mapping/Flowcharting

Process mapping/flowcharting is a graphical representation of workflow giving a clear understanding of a process or services of parallel processes. Process mapping/flowcharting are techniques that can be employed to not only procedure visual representation of the production processes but processes related to other departments.

FIGURE 4.28 illustrates process mapping/flowcharting diagram for approval of variation order.

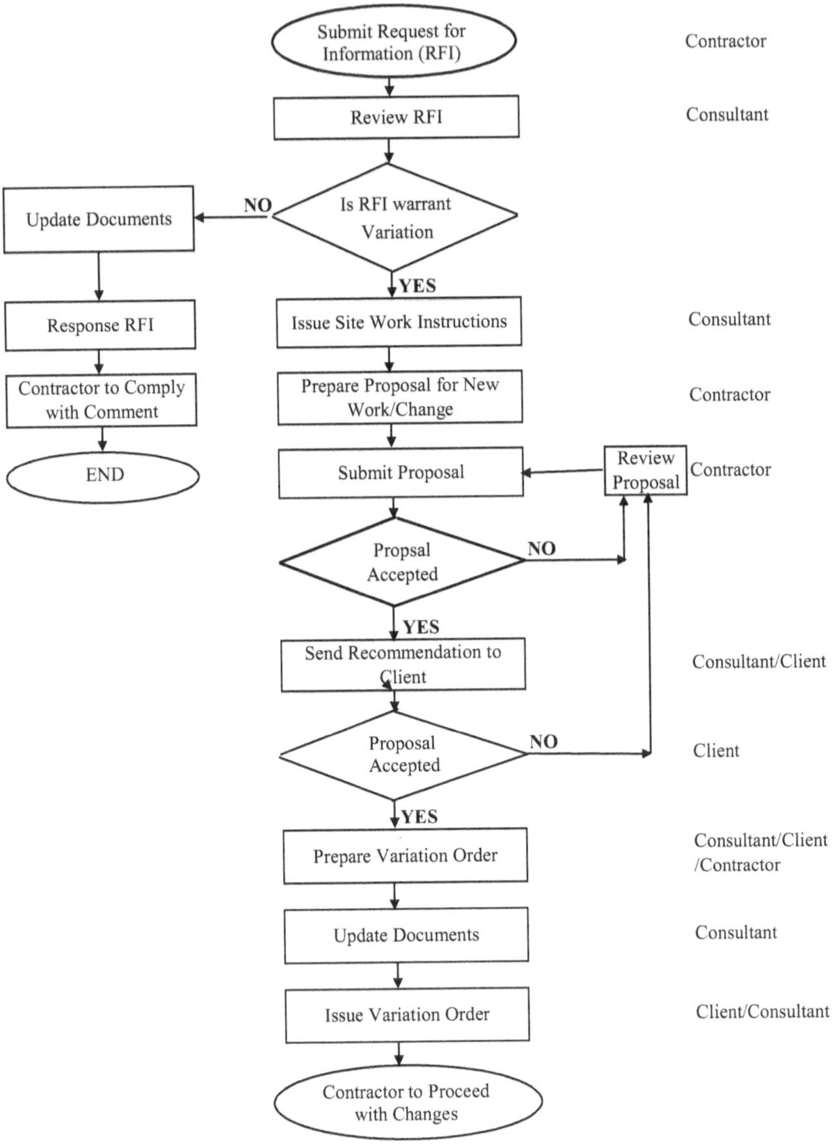

FIGURE 4.28 Process mapping/flowcharting.

4.2.4 Process Improvement Tools

TABLE 4.8 illustrates process improvement tools.

4.2.4.1 Root Cause Analysis

It is used to analyze the root causes of problems. The analysis is generally performed by using the Ishikawa diagram or cause and effect diagram.

FIGURE 4.29 illustrates root cause analysis for the rejection of executed/installed marble work.

4.2.4.2 Plan-Do-Check-Act Cycle

Plan-Do-Check-Act (PDCA) is mainly used for continuous improvement. It consists of a four-step model for carrying changes. The PDCA cycle model can be developed as a process improvement tool to reduce the cost of quality.

FIGURE 4.30 illustrates the PDCA cycle for preparation of shop drawings.

4.2.4.3 SIPOC Analysis

It is used to identify the Supplier-Input-Process-Output-Customer (SIPOC) relationship. The purpose of SIPOC analysis is to show the process flow by defining and documenting the suppliers, inputs, process steps, outputs, and customers.

TABLE 4.9 illustrates SIPOC Analysis for electrical panels.

4.2.4.4 Six Sigma—DMAIC

Six Sigma is discussed in detail under Section 4.2.9 of this chapter.

4.2.4.5 Failure Mode and Effects Analysis (FMEA)

Failure mode and effects analysis are also used as a process improvement tool. It identifies all the possible failures in the design of a product, process, or service and their effects on the product, process, and service. Its aim is to reduce the risk of failure and improve the process. Refer FIGURE 4.26 under Section 4.2.3.5 illustrates the failure mode and effects analysis process.

TABLE 4.8
Process Improvement Tools

Sr. No.	Name of Quality Tool	Usage
Tool 1	Root Cause Analysis	To identify root causes that caused the problem to occur.
Tool 2	PDCA Cycle	Used to plan for improvement followed by putting into action.
Tool 3	SIPOC Analysis	Used to identify supplier-input-process-output-customer relationships.
Tool 4	Six Sigma DMAIC	Used as an analytic tool for improvement.
Tool 5	Failure Mode and Effects Analysis (FMEA)	To identify and classify failures according to effect and prevent or reduce failure.
Tool 6	Statistical Process Control	Used to study how the process changes over time.

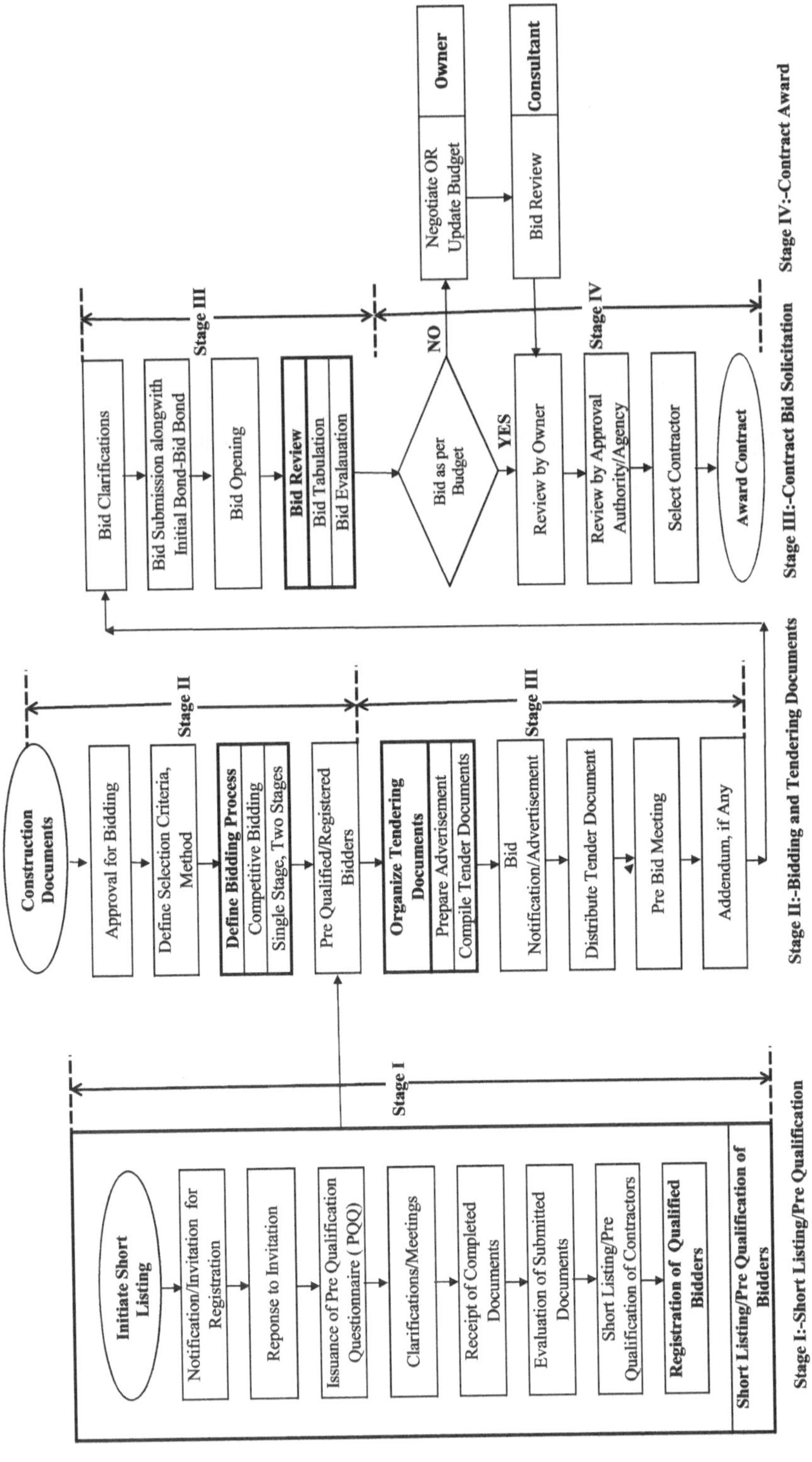

FIGURE 4.29 Root cause analysis.

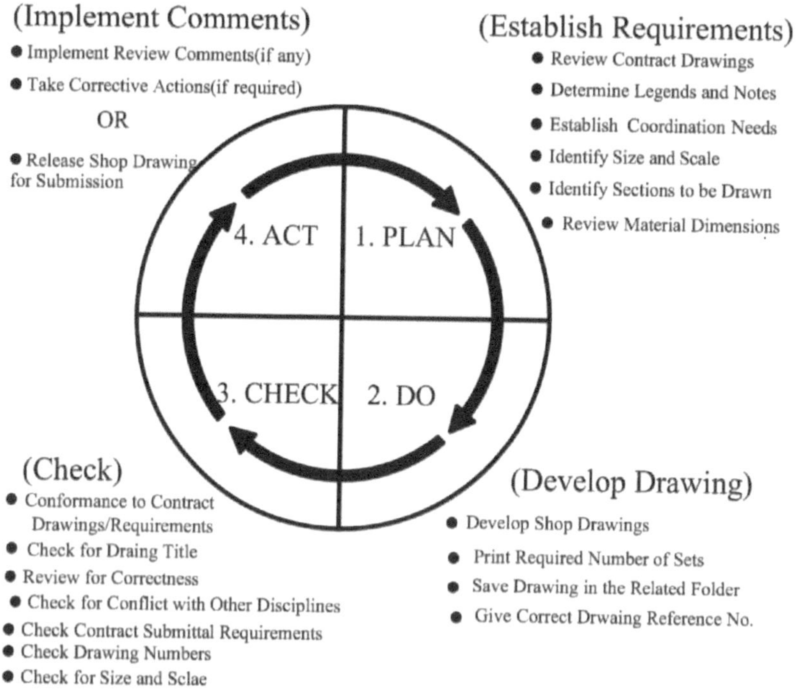

FIGURE 4.30 PDCA cycle for preparation of shop drawings.

TABLE 4.9
SIPOC Analysis

(Who are Suppliers)	(What the Suppliers are Providing)	(What is the Process)	(What is the Output of Process)	(Who are the Customers)
Supplier	**Inputs**	**Process**	**Outputs**	**Customer**
Electrical Panel Builder/ Assembler	Main Low-Tension Panel Main Switch Boards Distribution Boards Starter Panels Control Panels	Electrical Installation Work	Electrical Distribution Network	Power Supply for Project

4.2.4.6 Statistical Process Control

Statistical Process Control (SPC) is a quantitative approach based on the measurement of process control. Dr. Walter A. Shewhart developed the control charts as early as 1924. Statistical process control charts are used for the identification of common causes and special (or assignable) cause variations and for assisting in the diagnosis of quality problems. SPC charts reveal whether a process is "in control" stable and exhibiting only random variation or "out of control" and needing attention. A control chart is one of the key tools of SPC. It is used to monitor processes that are not in control, using measured ranges. There are two types of process control charts. These are:

1. Variable charts.
2. Attributes charts.

Variable charts relate to variable measurements such as length, width, temperature, weight, etc.

Attributes charts relate to the characteristics possessed (or not possessed) by the process or the product.

FIGURE 4.31 illustrates statistical control process charts for generator frequency.

4.2.5 Innovation and Creative Tools

TABLE 4.10 illustrates innovation and creative tools.

4.2.5.1 Brainstorming

Brainstorming is listing all the ideas put forth by a group in response to a given question or problem. It is a process of creating ideas by storming some objectives. In 1939, a team led by advertising executive Alex Osborn coined the term "brainstorm." According to Osborn, brainstorm means using the brain to storm a creative problem. Classical brainstorming is the most well-known and often-used technique for idea generation in a short period of time. It is based on the fundamental principles of deferment of judgment and that quantity breeds quality. It involves questions such as:

- Does the item have any design features that are not necessary?
- Can two or more parts be combined together?

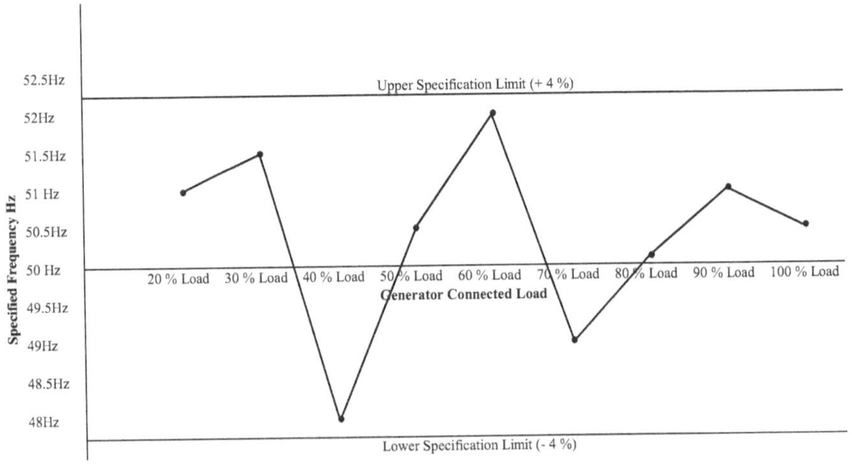

FIGURE 4.31 Statistical process control.

TABLE 4.10
Innovation and Creation Tools

Sr. No.	Name of Quality Tool	Usage
Tool 1	Brainstorming	Used to generate multiple ideas.
Tool 2	Delphi Technique	Used to get ideas from a select group of experts.
Tool 3	5W2H	The questions are used to understand why things happen the way they do.
Tool 4	Mind Mapping	Used to create a visual representation of many issues that can help get more understanding of the situation.
Tool 5	Nominal Group Technique	Used to enhance brainstorming by ranking the most useful ideas.
Tool 6	Six Sigma DMADV	Used primarily for the invention and innovation of modified or new products, services, or processes.
Tool 7	TRIZ	Used to provide systematic methods and tools for analysis and innovative problem-solving.

- How can we cut down the weight?
- Are these nonstandard parts that can be eliminated?

There are four rules for successful brainstorming:

1. Criticism is ruled out.
2. Freewheeling is welcomed.
3. Quantity is wanted.
4. Contribution and improvement are sought.

A classical brainstorming session has the following basic steps:

- *Preparation*: The participants are selected, and a preliminary statement of the problem is circulated.
- *Brainstorming*: A warm-up session with simple unrelated problems is conducted, the relevant problem and the four rules of brainstorming are presented, and ideas are generated and recorded using checklists and other techniques if necessary.
- *Evaluation*: The ideas are evaluated relative to the problem.

Generally, a brainstorming group should consist of four to seven people, although some suggest a larger group.

FIGURE 4.32 illustrates the brainstorming process.

4.2.5.2 Delphi Technique

The Delphi technique is intended to determine a consensus among experts on a subject matter. The goal of the Delphi technique is to pick the brains of experts in the

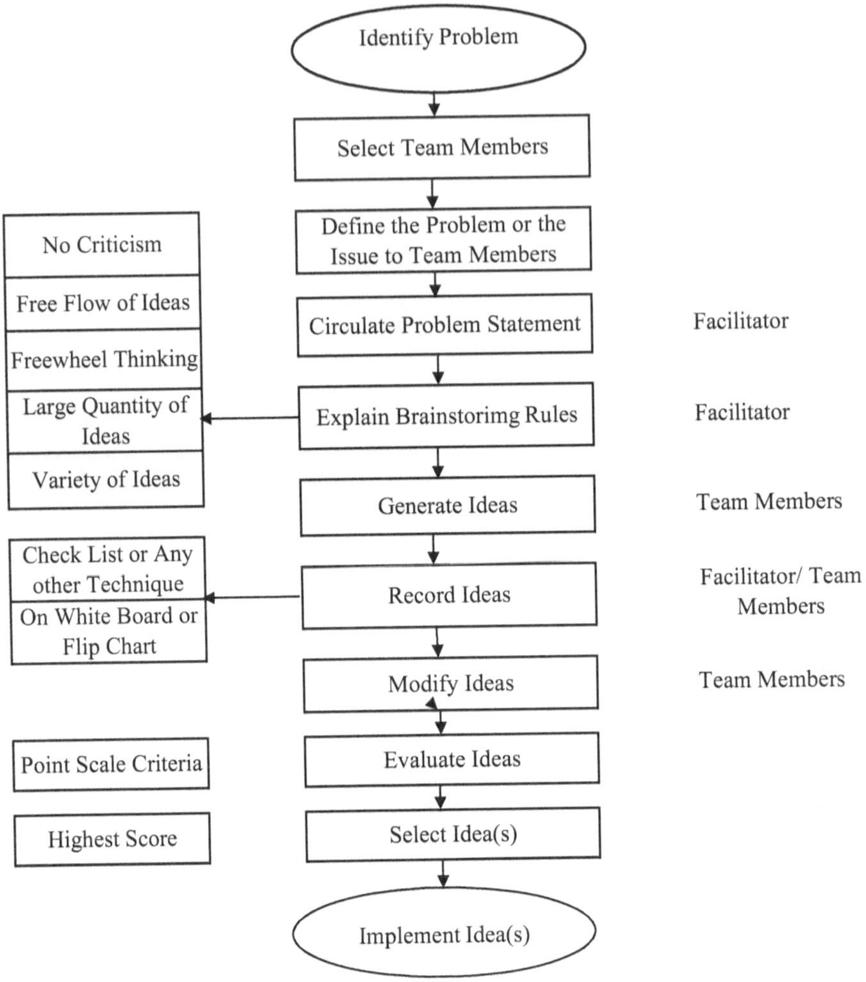

FIGURE 4.32 Brainstorming.

subject area and treat them as contributors to create ideas. It is a measure and method for consensus building by using questionnaires and obtaining responses from the panel of experts in the selected subjects. The Delphi technique employs multiple iterations designed to develop consensus opinions about a specific subject. The selected expert group answers questions by a facilitator. The responses are summarized and further circulated for group comments to reach consensus. The iteration/feedback process allows the team members to reassess their initial judgment and change or modify the earlier suggestions.

FIGURE 4.33 illustrates the Delphi technique process.

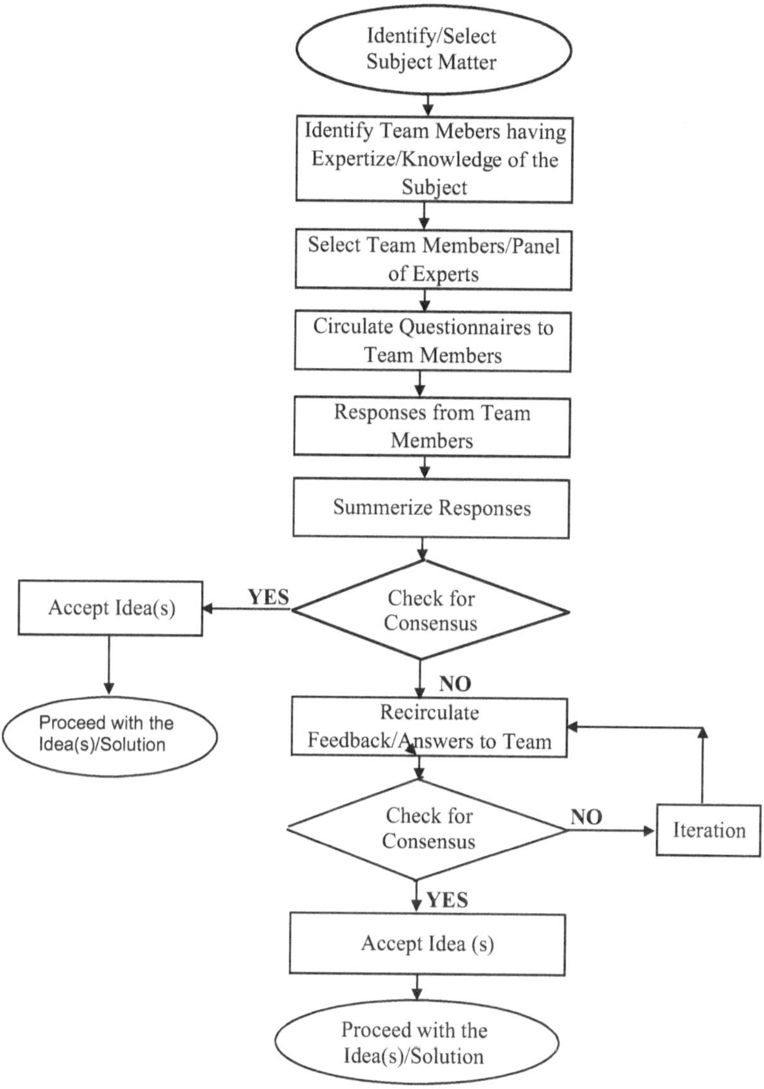

FIGURE 4.33 Delphi technique.

4.2.5.3 5W2H

5W2H is also used as an innovation and creative tool. The 5W2H is about asking questions to understand a process or problem.

The five Ws are:

1. Why?
2. What?

3. When?
4. Where?
5. Who?

And two Hs are:

1. How many?
2. How much

TABLE 4.11 illustrates 5W2H for the development of a new product.

4.2.5.4 Mind Mapping

Mind mapping is a graphical representation of ideas that can help get more understanding of the situation and create the solution or improve the task. FIGURE 4.34 illustrates a mind mapping sketch to improve site safety.

TABLE 4.11
5W2H

Serial Number	Why	Related Analyzing Question
1	Why	Why a new product?
2	What	What advantage will it have over other similar products?
3	Who	Who will be the customers for this product?
4	Where	Where can we market the product?
5	When	When will the product be ready for sale?
6	How many	How many pieces will be produced/sold per year?
7	How much	How much market share we will get for this product?

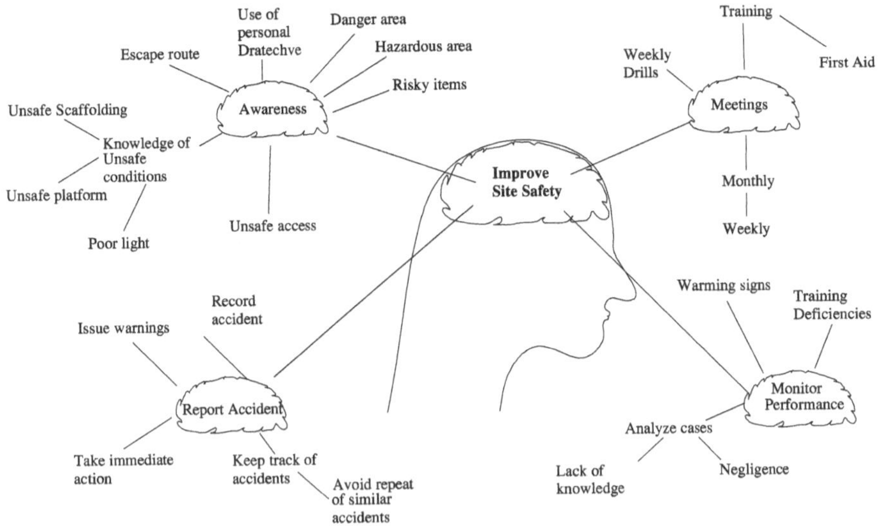

FIGURE 4.34 Mind mapping.

4.2.5.5 Nominal Group Technique

The Nominal Group Technique (NGT) involves a structural group meeting designed to incorporate individual ideas and judgments into a group consensus. By correctly applying the NGT, it is possible for groups of people (preferably five to ten) to generate alternatives or other ideas for improving the competitiveness of the firm. The technique can be used to obtain group thinking (consensus) on a wide range of topics. The technique, when properly applied, draws on the creativity of the individual participants while reducing two undesirable effects of most group meetings:

1. The dominance of one or more participants, and
2. The suppression of conflict ideas.

The basic format of an NGT session is as follows:

- Individual silent generation of ideas.
- Individual round-robin feedback and recording the ideas.
- Group's clarification of each idea.
- Individual voting and ranking to prioritize ideas.
- Discussion of group consensus results.

The NGT session begins with an explanation of the procedure and a statement of question(s), preferably written by the facilitator.

4.2.5.6 Six Sigma- DMADV

Six Sigma is discussed in detail under Section 4.2.9 of this chapter.

4.2.5.7 TRIZ

TRIZ is discussed in detail under Section 4.2.10 of this chapter.

4.2.6 Lean Tools

TABLE 4.12 illustrates Lean tools.

4.2.6.1 Cellular Design

A self-contained unit dedicated to performing all the operational requirements to accomplish sequential processing. With cellular design, individual cells can be fabricated and assembled to give the same performance and save time.

An electrical main switchboard may consist of a number of cells assembled together to perform desired operations. It helps easy maneuvering and assembling at the workplace for proper functioning.

FIGURE 4.35 illustrates the cellular design for the electrical panel.

4.2.6.2 Concurrent Engineering

The product life cycle begins with need and extends through concept design, preliminary design, detailed design, production or construction, product use, phase out, and disposal. Concurrent engineering is defined as a systematic approach to create

TABLE 4.12
Lean Tools

Sr. No.	Name of Quality Tool	Usage
Tool 1	Cellular Design	A self-contained unit dedicated to performing all the operational requirements to accomplish sequential processing.
Tool 2	Concurrent Engineering	It is used for product cycle reduction time. It is a systematic approach of creating a product design that simultaneously considers all elements of product life cycle.
Tool 3	5S	Used to eliminate waste that results from improper organization of work area.
Tool 4	Just in Time (JIT)	Used to reduce inventory levels, improve cash flow, and reduce space requirements for storage of material.
Tool 5	Kanban	Used to signal that more material is required to be ordered. It is used to eliminate waste from inventory.
Tool 6	Kaizen	Used for continually eliminating waste from manufacturing processes by combining the collective talent of the company.
Tool 7	Mistake Proofing	Used to eliminate the opportunity for error by detecting the potential source of error.
Tool 8	Outsourcing	It is contracting out certain works, processes, and services to specialists in the discipline area.
Tool 9	Poka-Yoke	Used to detect the abnormality or error and fix or correct the error and take action to prevent the error.
Tool 10	Single Minute Exchange of Die (SMED)	Used to reduce set up time for change over to new process.
Tool 11	Value Stream Mapping	Used to establish flow of material or information and eliminate waste and add value.
Tool 12	Visual Management	It addresses both visual display and control. It exposes waste elimination/prevention.
Tool 13	Waste Reduction	It focuses on reducing waste.

a product design that simultaneously considers all the elements of the product life cycle thus reducing the product life cycle time. It is used to expedite the development and launch of new products. In construction projects, construction can simultaneously start while the design is under development.

FIGURE 4.36 illustrates concurrent engineering for construction project life cycle.

4.2.6.3 5S

5S is a systematic approach for the improvement of quality and safety by organizing a workplace. It is a methodology that advocates,

- What should be kept?
- Where should it be kept?
- How should it be kept?

5S is a Japanese concept of housekeeping having reference to five Japanese words starting with the letter "S." TABLE 4.13 illustrates 5S for construction projects.

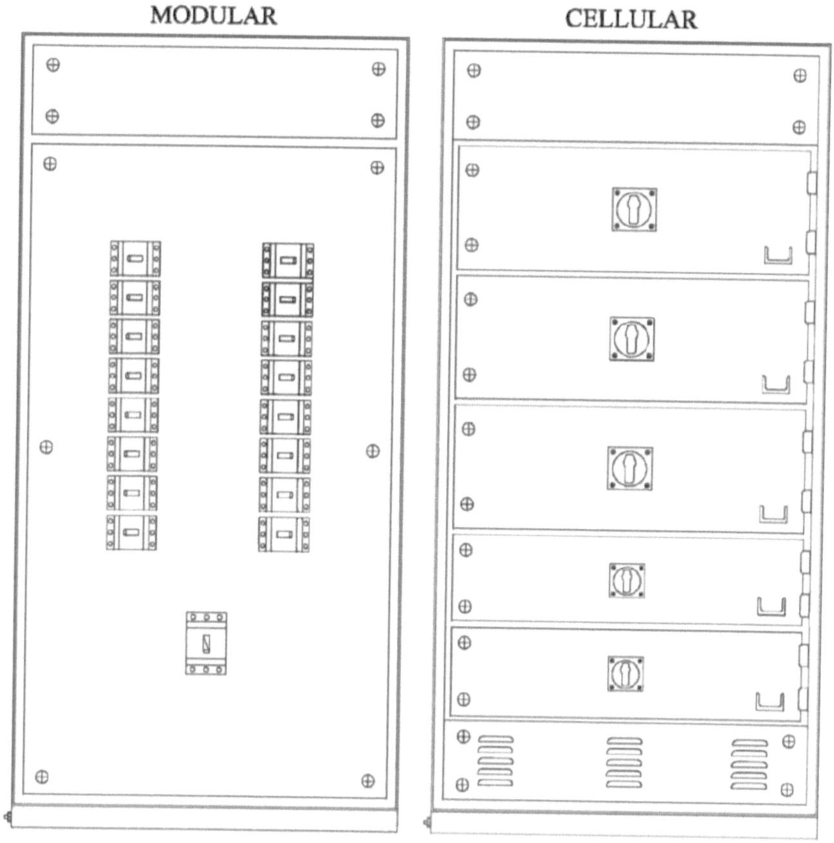

FIGURE 4.35 Cellular design.

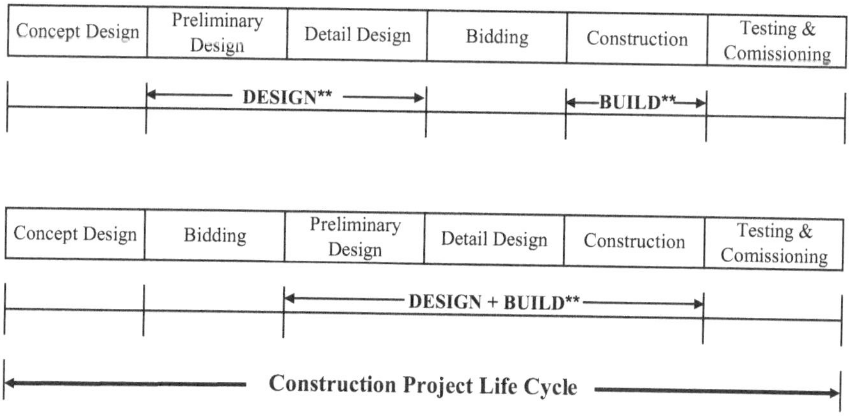

FIGURE 4.36 Concurrent engineering.

TABLE 4.13
5S for Construction

Sr. No.	5S	Related Action
1	Sort	• Determine what is to be kept in the open and what is under shed. • Allocate an area for each type of construction equipment and machinery. • Allocate an area for electrical tools. • Allocate an area for hand tools. • Allocate an area for construction material/equipment to be used/installed in the project. • Allocate an area for hazardous, flammable materials. • Allocate an area for chemicals, paints. • Allocate an area for spare parts for maintenance.
2	Set in Order	• Keep/arrange equipment in such a way that their maneuvering/movement shall be easy. • Vehicles to be parked in the yard in such a way that frequently used vehicles are parked near the gate. • Frequently used equipment/machinery to be located near the workplace. • Set boundaries for different types of equipment and machinery. • Identify and arrange tools for easy access. • Identify and store material/equipment as per relevant division/section of contract documents. • Identify and store materials in accordance with their usage as per construction schedule. • Determine items that need special conditions. • Mark/tag the items/material. • Display route map and location. • Put the material in sequence as per their use. • Frequently used consumables to be kept near workplace. • Label on the drawer with list of contents. • Keep shuttering material in one place. • Determine inventory level of consumable items.

3	Sweeping	• Clean site on daily basis by removing:
		  • cut pieces of reinforced bars
		  • cut pieces of plywood
		  • left out concrete
		  • cut pieces of pipes
		  • cut pieces of cables and wires
		  • used welding rods
		  • clean equipment and vehicles
		  • check electrical tools after return by the technician
		  • attend to breakdown report

4	Standardize	• Standardize the store by allocating separate areas for material used by different divisions/sections. • Standardize area for long lead items. • Determine regular schedule for cleaning the workplace. • Make available standard tool kit/box for a group of technicians. • Make everyone informed of their responsibilities and related area where the things are to be placed and are available. • Standardize the store for consumable items. • Inform suppliers/vendors in advance the place for delivery of material.
5	Sustain	• Follow the system till the end of the project.

4.2.6.4 Just in Time (JIT)

It is used to reduce inventory levels, improve cash flow, and reduce storage space requirements for materials. For example,

1. Concrete block can be received at the site just before the start of block work and can be stacked near the work area where masonry work is in progress.
2. The chiller can be received at the site and directly placed on the chiller foundation without storing it in the storage yard.

4.2.6.5 Kanban

It is used to signal more material is required to be ordered. It is used to eliminate waste from inventory and inventory control thus to avoid extra storage required for large inventory. In construction projects, electrical wires for circuiting can be ordered to be received at the site when the wire-pulling work is in progress. Similarly, concrete blocks and false ceiling tiles can be ordered and received as and when required.

4.2.6.6 Kaizen

It is used for continually improving through small changes to eliminate waste from the manufacturing process by combining the collective talent of every employee of the company.

4.2.6.7 Mistake Proofing

Mistake proofing is used to eliminate the opportunity for error by detecting the potential source of error. Mistakes are generally categorized as follows:

- Information
- Mismanagement
- Omission
- Selection

TABLE 4.14 illustrates a mistake-proofing chart for eliminating design errors.

4.2.6.8 Outsourcing

It is contracting out certain works, processes, and services to specialists in a particular discipline area. For example, in construction projects, the following is a list of some of the work which are outsourced (sub-contracted):

- Structural concrete.
- Waterproofing work.
- HVAC work.
- Fire suppression work.
- Water supply piping.
- Electrical work.
- Security system.
- Low voltage work.
- Landscape work.

TABLE 4.14
Mistake Proofing

Serial Number	Items	Points to be Considered to Avoid Mistakes
1	Information	1. Terms of reference (TOR). 2. Client's preferred requirements matrix. 3. Data collection. 4. Regulatory requirements. 5. Codes and standards. 6. Historical data. 7. Organizational requirements.
2	Mismanagement	1. Compare production with actual requirements. 2. Interdisciplinary coordination. 3. Application of different codes and standards. 4. Drawing sizes of different trades/specialist consultants.
3	Omission	1. Review and check design with TOR. 2. Review and check design with client requirements. 3. Review and check design with Regulatory requirements. 4. Review and check design with codes and standards. 5. Check for all required documents.
4	Selection	1. Qualified team members. 2. Available material. 3. Installation methods.

4.2.6.9 Poka-Yoke

Poka-Yoke is a quality management concept developed by Shigeo Shino to prevent human errors from occurring in the production line. The main objective of Poka-Yoke is to achieve zero defects.

4.2.6.10 Single-Minute Exchange of Die (SMED)

SMED is used to reduce set-up time for change over to a new process.

For example, a spare circuit breaker of a similar rating can be used as an immediate replacement for a damaged circuit breaker in the electrical distribution board to avoid the breakdown of the electrical supply for a long duration. Subsequently, a new circuit breaker can be affixed in place of a spare breaker.

4.2.6.11 Value Stream Mapping

It is used to establish the flow of material or information and eliminate waste and add value. Value stream mapping is used to identify areas for improvement.

FIGURE 4.37 illustrates a value stream mapping diagram for an emergency power system.

4.2.6.12 Visual Management

It addresses both visual display and control. It exposes waste elimination/prevention. Visual displays present information, while visual control focuses on a need to act.

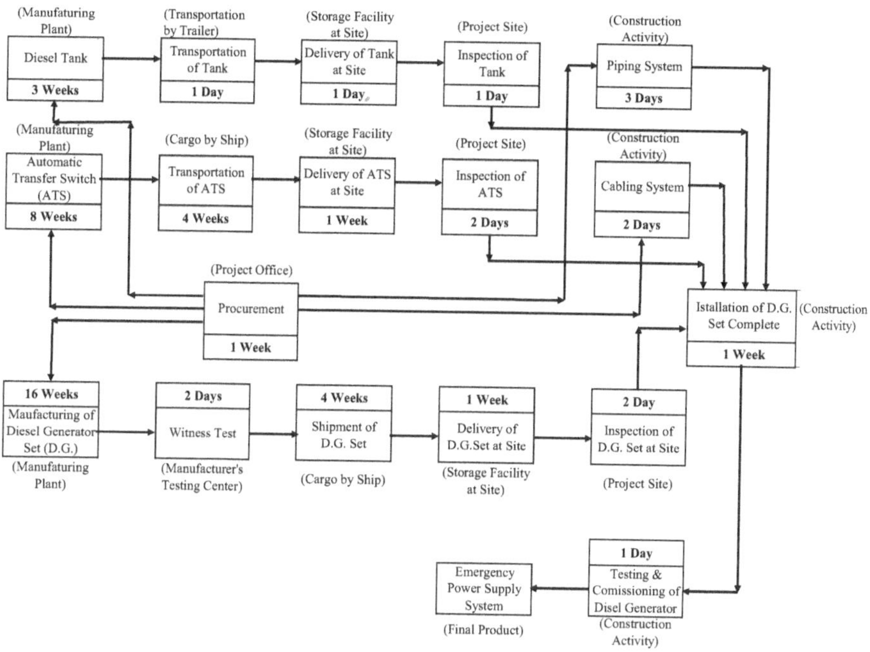

FIGURE 4.37 Value stream mapping.

4.2.6.13 Waste Reduction

It focuses on reducing waste. The following are the general types of waste:

1. Defective parts.
2. Delays, waiting.
3. Excess inventory.
4. Misuse of resources.
5. Overproduction.
6. Processing.
7. Transportation.
8. Untapped resources.
9. Wasted motion.

4.2.7 Cost of Quality

4.2.7.1 Introduction

Quality has an impact on the costs of products and services. The cost of poor quality is the annual monetary loss of products and processes that are not achieving their quality objective. The main components of the cost of low quality are:

1. Cost of conformance.
2. Cost of non-conformance.

TABLE 4.15
Elements of Cost of Quality

Cost of Compliance	Cost of Noncompliance
• Quality planning.	• Scrap.
• Process control planning.	• Rework.
• Quality training.	• Corrective action.
• Quality audit.	• Additional material/inventory cost.
• Design review.	• Expedition.
• Product design validation.	• Customer complaints.
• Work procedure.	• Product recalls.
• Method statement.	• Warranty.
• Process validation.	• Maintenance service.
• Field testing.	• Field repairs.
• Third-party inspection.	• Rectification of returned material.
• Receiving inspection.	• Re inspection or re-test.
• Prevention action.	• Downgrading.
• In-process inspection.	• Loss of business.
• Outside endorsement.	
• Calibration of equipment.	
• Laboratory acceptance testing.	

TABLE 4.15 illustrates elements of cost of quality.

4.2.7.2 Categories of Costs

Costs of poor quality are the costs associated with providing poor-quality products or services. These are incurred because of poor quality, costs that would not be incurred if things were done right from the time and at every time to follow thereafter in order to achieve the quality objective. There are four categories of costs:

1. Internal failure costs—The costs associated with defects found before the customer receives the product or service. It also consists of cost of failure to meet customer satisfaction and needs and cost of inefficient processes.
2. External failure costs—The cost associated with defects found after the customer receives the product or service. It also includes lost opportunities for sales revenue.
3. Appraisal costs—The costs incurred to determine the degree of conformance to quality requirements.
4. Prevention costs—The costs incurred to keep failure and appraisal costs to a minimum.

These cost categories allow the use of quality cost data for a variety of purposes. Quality costs can be used for measurement of progress, for analyzing the problem, or for budgeting. By analyzing the relative size of the cost categories, the company can determine if its resources are properly allocated.

4.2.7.3 Quality Cost in Construction

Quality of construction is defined as:

1. Scope of Work
2. Time
3. Budget

Cost of quality refers to the total cost incurred during the entire life cycle of a construction project in preventing nonconformance to owner requirements (defined scope). There are certain hidden costs that may not affect directly the overall cost of the project; however, it may cost the consultant/designer to complete the design within the stipulated schedule to meet owner requirements and conformance to all the regulatory codes/standards, and for a contractor to construct the project within stipulated schedule meeting all the contract requirements. Rejection /non-approval of executed/installed work by the supervisor due to noncompliance with the specification will cause the contractor loss in terms of:

- Material
- Manpower
- Time

The contractor will have to rework or rectify the work which will need additional resources and will need extra time to do the work as specified. This may disturb the contractor's work schedule and affect the execution of other activities. The contractor has to emphasize upon "Zero Defect" policy, particularly for concrete works. To avoid rejection of work, the contractor has to take the following measures:

1. Execution of work as per approved shop drawings using approved material.
2. Following the approved method of statement or manufacturer-recommended method of installation.
3. Conduct continuous inspection during the construction/installation process.
4. Employ a properly trained workforce.
5. Maintain good workmanship.
6. Identify and correct deficiencies before submitting the checklist for inspection and approval of work.
7. Coordinate requirements of other trades, for example, if any opening is required in the concrete beam for the crossing of services pipe.

Timely completion of the project is one of the objectives to be achieved. To avoid delay in the completion schedule, proper planning and scheduling of construction activities is necessary. Since construction projects involve many participants, it is essential that the requirements of all the participants are fully coordinated. This will ensure the execution of activities as planned and result in the timely completion of the project.

Normally the construction budget is fixed at the inception of the project, therefore it is required to avoid variations during the construction process as it may take time to get approval of additional budget resulting in time extension to the project. Quality costs related to construction projects can be summarized as follows:

Internal Failure Costs:

- Rework
- Rectification
- Rejection of Checklist
- Corrective Action

External Failure Costs:

- Breakdown of Installed System
- Repairs
- Maintenance
- Warranty

Appraisal Costs:

- Design review/preparation of shop drawings.
- Preparation of composite/coordination drawings.
- On-site material inspection/test.
- Off-site material inspection/test.
- Pre-checklist inspection.

Prevention Costs:

- Preventive Action
- Training
- Work Procedures
- Method Statement
- Calibration of Instruments/Equipment

4.2.8 Quality Function Deployment (QFD)

Quality Function Deployment (QFD) is a technique to translate customer requirements into technical requirements. It was developed in Japan by Dr. Yoji Akao in the 1960s to transfer the concepts of quality control from the manufacturing process into the new product development process. QFD is referred to as the "Voice of Customer" which helps in identifying and developing customer requirements through each stage of product or service development. It is a development process that utilizes a comprehensive matrix involving project team members.

QFD involves constructing one or more matrices containing information related to others. The assembly of several matrices showing the correlation with one another is called "the house of quality." The "house of quality" matrix is the most recognized form of QFD. QFD is being applied virtually in every industry and business from aerospace, communication, software, transportation, manufacturing, services industry, and construction industry. The house of quality is made up of the following major components:

1. What's
2. How's
3. Correlation Matrix (Roof)—Technical requirements
4. Interrelationship Matrix
5. Target Value
6. Competitive Evaluation

What is the first step in developing the house of quality. It is a structured set of needs/requirements ranked in terms of priority and the levels of importance being specified quantitatively. It is generated by using questions such as:

- What are the types of finishes needed for the building?
- What type of air conditioning system is required for the building?
- What type of communication system is required for the building?
- What type of flooring material is required?
- Whether the building needs any security system?

How is the second step in which the project team members translate the requirements (What) into technical design characteristics (Specifications) and are listed across the columns.

The correlation matrix identifies technical interaction or physical relationships among the technical specifications.

The interrelationship matrix illustrates team members' perception of the interrelationship between the owner's requirements and technical specifications.

The bottom part allows for technical comparison between possible alternatives, target values for each technical design characteristic, and performance measurement.

The right side of the house of quality is used for planning purposes. It illustrates customer perceptions observed in the market survey.

QFD technique can be used to translate the owner's needs/requirements into the development of a set of technical requirements during conceptual design.

FIGURE 4.38 illustrates the house of quality for a hospital building.

4.2.9 Six Sigma

4.2.9.1 Introduction

Six Sigma is, basically, a process quality goal. It is a process quality technique that focuses on reducing variation in the process and preventing deficiencies in the product. In a process that has achieved Six Sigma capability, the variation is small compared to the specification limits.

Sigma is a Greek letter σ and stands for standard deviation. Standard deviation is a statistical way to describe how much variation exists in a set of data, a group of

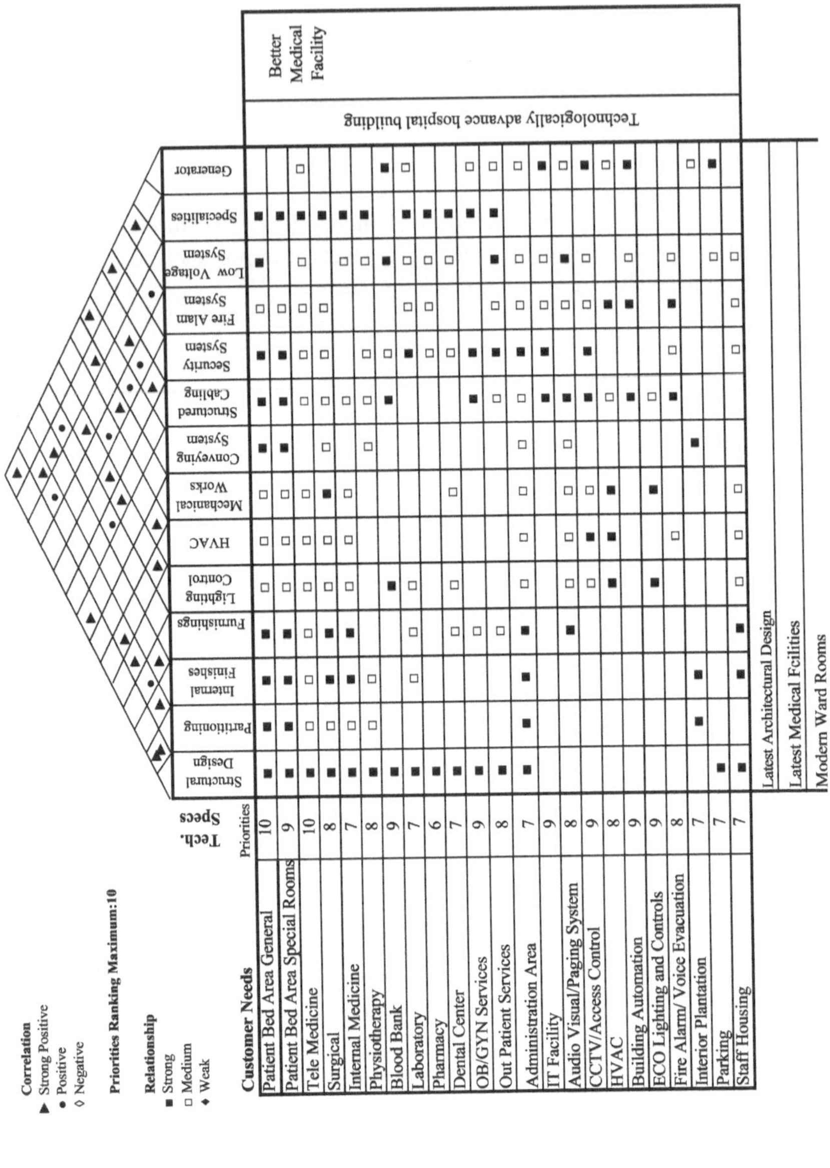

FIGURE 4.38 House of quality for hospital building.

items, or a process. Standard deviation is the most useful measure of dispersion. Six Sigma means that for a process to be capable at Six Sigma level, the specification limits should be at least 6 σ from the average point. So, the total spread between the upper specification (control) limit and lower specification (control) limit should be 12 σ. With Motorola's Six Sigma program no more than 3.4 defects per million fall outside specification limits with the process shift of not more than 1.5 σ from the average or mean. Six Sigma started as a defect reduction effort in manufacturing and was then applied to other business processes for the same purpose.

Six Sigma is a measurement of "goodness" using a universal measurement scale. Six Sigma provides a relative way to measure improvement. Universal means sigma can measure anything from coffee mug defects to missed chances to close a sales deal. It simply measures how many times a customer's requirements were not met (a defect), given a million opportunities. Sigma is measured in defects per million opportunities (DPMO). For example, a level of sigma can indicate how many defective coffee mugs were produced when one million were manufactured. Levels of sigma are associated with improved levels of goodness. To reach a level of Three Sigma, you can only have 66,811 defects, given a million opportunities. A level of Five Sigma only allows 233 defects. Minimizing variation is a key focus of Six Sigma. Variation leads to defects, and defects lead to unhappy customers. To keep customers satisfied, loyal, and coming back, you have to eliminate the sources of variation. Whenever a product is created or a service performed, it needs to be done the same way every time, no matter who is involved. Only then will you truly satisfy the customer. FIGURE 4.39 illustrates the Six Sigma roadmap.

4.2.9.2 Six Sigma Methodology

Six Sigma is an overall business improvement methodology that focuses an organization on:

- Understanding and managing customer requirements.
- Aligning key business processes to achieve these requirements.
- Utilizing rigorous data analysis to minimize variation in these processes.
- Driving rapid and sustainable improvement in business processes by reducing defects, cycle time, impact to the environment, and other undesirable variations.
- Timely execution.

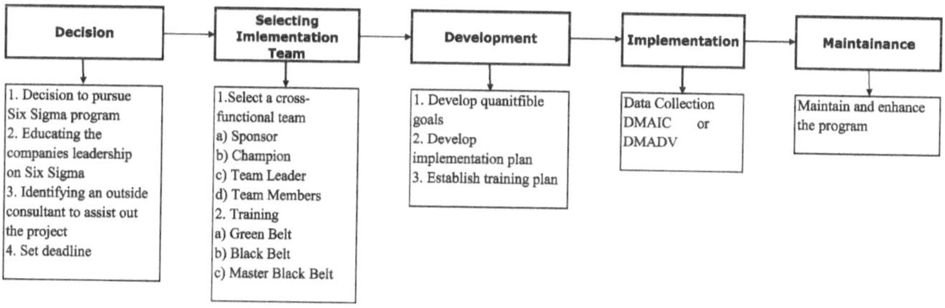

FIGURE 4.39 Six Sigma roadmap.

As a management system, Six Sigma is a high-performance system for executing business strategy. It uses the concept of fact and data to drive better solutions. Six Sigma is a top-down solution to help organizations:

- Align their business strategy to critical improvement efforts.
- Mobilize teams to attack high-impact projects.
- Accelerate improved business results.
- Govern efforts to ensure improvements are sustained.

Six Sigma methodology also focuses on:

- Leadership Principles
- Integrated Approach to Improvement
- Engaged Teams
- Analytic Tool
- Hard Coded Improvements

4.2.9.2.1 Leadership Principles

Six Sigma methodology has four leadership principles. These are:

1. Align
2. Mobilize
3. Accelerate
4. Govern

Brief description of these leadership principles is:

1. ALIGN—Leadership should ensure that all improvement projects are in line with the organization's strategic goals.

Alignment begins with the leadership team developing a scorecard. This vital tool, the cornerstone of the Six Sigma business improvement campaign, translates strategy into tactical operating terms. The scorecard also defines metrics an organization can use to determine success. Just as a scoreboard at a sporting event tells you who is winning, the scorecard tells leadership how well the company is meeting its goals.

2. MOBILIZE—Leadership should enable teams to take action by providing clear direction, feasible scope, a definition of success, and rigorous reviews.

Mobilizing sets clear boundaries, lets people go to work, and trains them as required.

The key to mobilizing is focus—lack of focused action was one of the downfalls of previous business improvement efforts. True focus means the project is correctly aligned with the organization's scorecard. Mobilized teams have a valid reason for engaging in improvement efforts—they can see benefits for the customer. The project has strategic importance, and they know it. They know exactly what must be done and the criteria they can use to determine success.

3. ACCELERATE—Leadership should drive a project to rapid results through tight clock management, training as needed, and shorter deadlines.

More than 70% of all improvement initiatives fail to achieve desired results in time to make a difference. For projects to make an impact, they must achieve results quickly, and that is what acceleration is all about.

Accelerate leadership principle involves three main components:

a. Action learning.
b. Clock management.
c. Effective planning.

Accelerate employs "action learning" methodology to quickly bridge from "learning" to "doing." Action learning mixes traditional training with direct application. Training is received while working on a real-world project, allowing plenty of opportunities to apply new knowledge. The instructor is not simply a trainer, but a coach as well, helping work with real-world projects. Action learning accelerates improvement over traditional learning methods. It helps with receiving training and also completing a worthwhile project at the same time. In addition to the four-to-six-month time frame, acceleration requires teams to set deadlines that are reinforced through rigorous reviews.

4. GOVERN Leadership must visibly sponsor projects and conduct regular and rigorous reviews to make critical mid-course corrections.

The fourth leadership principle is Govern. Once leadership selects an improvement opportunity, their work is not done. They must remain ultimately responsible for the success of that project. Govern requires leaders to drive for results.

While governing a Six Sigma project, you need:

- A regular communications plan and a clear review process.
- Actively sponsor teams and their projects.
- To encourage proactive dialogue and knowledge-sharing on the team and throughout the organization.

4.2.9.2.2 Six Sigma Team

Teamwork is absolutely vital for complex Six Sigma projects. For teams to be effective, they must be engaged—involved, focused, and committed to meeting their goals. Engaged teams must have leadership support. There are four types of teams. These are:

1. Black Belts
2. Green Belts

3. Breakthrough
4. Blitz

A brief description of these teams is as follows:

4.2.9.2.2.1 Black Belt

Black Belt teams are led by a Black Belt and may have Green Belts and functional experts assigned to complex, high-impact process improvement projects or designing new products, services, or complex processes. Black Belts are internal Six Sigma practitioners, skilled in the application of rigorous statistical methodologies, and they are crucial to the success of Six Sigma. Their additional training and experience provide them with the skills they need to tackle difficult problems. Black Belts have many responsibilities.

- Function as a team leader on Black Belt projects.
- Integrate their functional discipline with statistical, project, and interpersonal skills.
- Serve as internal consultants.
- Tackle complex, high-impact improvement opportunities.
- Mentor and train Green Belts.

4.2.9.2.2.2 Green Belt

Led by a Green Belt and comprised of non-experts, Green Belt teams tackle less complex, high-impact process improvement projects. Green Belt teams are often coached by Black Belts or Master Black Belts.

Green Belts are also essential to the success of Six Sigma. They perform many of the same functions as Black Belts, but their work requires less complex analysis. Green Belts are trained in basic problem-solving skills and the statistical tools needed to work effectively as members of process improvement teams.

Green Belt responsibilities include:

- Acting as Team Leader on business improvements requiring less complex analysis.
- Adding their unique skills and experiences to the team.
- Working with the team to come up with inventive solutions.
- Performing basic statistical analysis.
- Conferring with a Black Belt as questions arise.

4.2.9.2.2.3 Breakthrough

While creating simple processes, sophisticated statistical tools may not be needed. Breakthrough teams are typically used to define low-complexity, new processes.

4.2.9.2.2.4 Blitz

Blitz teams are put in place to quickly execute improvements produced by other projects. These teams can also implement digitization for efficiency using a new analytic tool set.

For a typical Six Sigma project, four critical roles exist.

1. Sponsor
2. Champion
3. Team Leader
4. Team Member

A Sponsor typically:

- Remains ultimately accountable for a project's impact.
- Provides project resources.
- Reviews monthly and quarterly achievements, obstacles, and key actions.
- Supports the project Champion by removing barriers as necessary.

A Champion typically:

- Reviews weekly achievements, obstacles, and key actions.
- Meets with the team weekly to discuss progress.
- Reacts to changes in critical performance measures as needed.
- Supports the Team Leader, removing barriers as necessary.
- Helps ensure project alignment.

A Team Leader typically:

- Leads improvement projects through an assigned, disciplined methodology.
- Works with the Champion to develop the Team Charter, review project progress, obtain necessary resources, and remove obstacles.
- Identifies and develops key milestones, timelines, and metrics for improvement projects.
- Establishes weekly, monthly, and quarterly review plans to monitor team progress.
- Supports the work of team members as necessary.

Team members typically:

- Assist the Team Leader.
- Follow a disciplined methodology.
- Ensure the Team Charter and timeline are being met.
- Accept and execute assignments.
- Add their views, opinions, and ideas.

4.2.9.3 Analytic Tool Sets

The following are the analytic tools used in Six Sigma projects:

4.2.9.3.1 Ford's Global 8D Tool

What Problem Needs Solving? →
Who should help solve the problem? →

How do we quantify symptoms? →
How do we contain it? →
What is the root cause? →
What is the permanent corrective action? →
How do we implement? →
How can we prevent this in future? →
Who should we reward? →

Ford's Global 8D tool is primarily used to bring performance back to a previous level.

4.2.9.3.2 DMADV Tool Set Phases

Define → What is Important?
Measure → What is needed?
Analyze → How will we fulfill?
Design → How do we build it?
Verify → How do we know it will work?

DMADV tool is used primarily for the invention and innovation of modified or new products, services, or processes. Using this toolset, Black Belts optimize performance before production begins. DMADV is proactive, solving problems before they start. This tool is also called DFSS (Design for Six Sigma).

TABLE 4.16 lists the fundamental objectives of DMADV.

4.2.9.3.2.1 DMADV Process

DEFINE PHASE: What is important?
(Define the project goals and customer deliverables.)

TABLE 4.16
Fundamental Objectives of Six Sigma DMADV Tool

DMADV	Phase	Fundamental Objective
1	**Define**—What is important?	Define the project goals and customer deliverables (internal and external).
2	**Measure**—What is needed?	Measure and determine customer needs and specifications.
3	**Analyze**—How do we fulfill?	Analyze process options and prioritize based on capabilities to satisfy customer requirements.
4	**Design**—How do we build it?	Design detailed process(es) capable of satisfying customer requirements.
5	**Verify**—How do we know it will work?	Verify design performance capability.

Key deliverables of this phase are:

- Establish the goal.
- Identify the benefits.
- Select project team.
- Develop project plan.
- Project charter.

MEASURE PHASE: What is needed?
(Measure and determine customer needs and specifications.)
Key deliverables in this phase are:

- Identify specification requirements.

ANALYZE PHASE: How do we fulfill?
(Analyze process options and prioritize based on capability to satisfy customer requirements.)
Key deliverables in this phase are:

- Design generation (data collection).
- Design analysis.
- Risk analysis.
- Design model (prioritization of data under major variables).

DESIGN PHASE: How do we build it?
(Design detailed process(es) capable of satisfying customer requirements.)
Key deliverables in this phase are:

- Constructing a detailed design.
- Converting CTQs (critical to quality) into CTPs (critical to process elements).
- Estimate the capabilities of the CTPs in the design.
- Preparing a verification plan.

VERIFY PHASE: How do we know it will work?
(Verify design performance capability.)
Key deliverables in this phase are:

- Designing a control and transition plan.

4.2.9.3.3 DMAIC Tool

Define → What is Important?
Measure → How are we doing?
Analyze → What is wrong?
Improve → What needs to be done?
Control → How do we guarantee performance?

TABLE 4.17
Fundamental Objectives of Six Sigma DMAIC Tool

DMAIC	Phase	Fundamental Objective
1	**Define**—What is important?	Define the project goals and customer deliverables (internal and external).
2	**Measure**—How are we doing?	Measure the process to determine current performance.
3	**Analyze**—What is wrong?	Analyze and determine the root cause(es) of the defects.
4	**Improve**—What needs to be done?	Improve the process by permanently removing the defects.
5	**Control**—How do we guarantee performance?	Control the improved process's performance to ensure sustainable results.

DMAIC tool refers to a data-driven quality strategy and is used primarily for the improvement of an existing product, service, or process.

TABLE 4.17 lists the fundamental objectives of DMAIC.

4.2.9.3.3.1 The DMAIC Process

The majority of the time, Black and Green Belts approach their projects with the DMAIC analytic tool set, driving process performance to never-before-seen levels.

DMAIC has the following fundamental objectives:

1. Define Phase: Define the project and customer deliverables.
2. Measure Phase: Measure the process performance and determine current performance.
3. Analyze: Collect, analyze, and determine the root cause(s) of variation and process performance.
4. Improve: Improve the process by diminishing defects with alternative remedial.
5. Control: Control improved process performance.

The DMAIC process contains five distinct steps that provide a disciplined approach to improving existing processes and products through the effective integration of project management, problem-solving, and statistical tools. Each step has fundamental objectives and a set of key deliverables, so the team member will always know what is expected of him/her and his/her team.

DMAIC stands for the following:

- Define opportunities.
- Measure performance.
- Analyze opportunity.
- Improve performance.
- Control performance.

DEFINE OPPORTUNITIES (What is important?)
The objective of this phase is:

> To identify and/or validate the improvement opportunities that will achieve the organization's goals and provide the largest payoff, develop the business process, define critical customer requirements, and prepare to function as an effective project team.

Key deliverables in this phase include:

- Team charter.
- Action plan.
- Process map.
- Quick win opportunities.
- Critical customer requirements.
- Prepared team.

MEASURE PERFORMANCE (How are we doing?)
The objectives of this phase are:

- To identify critical measures that are necessary to evaluate the success or failure, meet critical customer requirements, and begin developing a methodology to effectively collect data to measure process performance.
- To understand the elements of the Six Sigma calculation and establish baseline sigma for the processes the team is analyzing.

Key deliverables in this phase include:

- Input, process, and output indicators.
- Operational definitions.
- Data collection format and plans.
- Baseline performance.
- Productive team atmosphere.

ANALYZE OPPORTUNITY (What is wrong?)
The objectives of this phase are:

- To stratify and analyze the opportunity to identify a specific problem and define an easily understood problem statement.
- To identify and validate the root causes and thus the problem the team is focused on.
- To determine true sources of variation and potential failure modes that lead to customer dissatisfaction.

Key deliverables in this phase include:

- Data analysis.
- Validated root causes.

- Sources of variation.
- Failure modes and effects analysis (FMEA).
- Problem statement.
- Potential solutions.

IMPROVE PERFORMANCE (What needs to be done?)

The objectives of this phase are:

- To identify, evaluate, and select the right improvement solutions.
- To develop a change management approach to assist the organization in adapting to the changes introduced through solution implementation.

Key deliverables in this phase include:

- Solutions.
- Process maps and documentation.
- Pilot results.
- Implementation milestones.
- Improvement impacts and benefits.
- Storyboard.
- Change plans.

CONTROL PERFORMANCE (How do we guarantee performance?)

The objectives of this phase are:

- To understand the importance of planning and executing against the plan and determine the approach to be taken to ensure achievement of the targeted results.
- To understand how to disseminate lessons learned, identify replication and standardization opportunities/processes, and develop related plans.

Key deliverables in this phase include:

- Process control systems.
- Standards and procedures.
- Training.
- Team evaluation.
- Change implementation plans.
- Potential problem analysis.
- Solution results.
- Success stories.
- Trained associates.
- Replication opportunities.
- Standardization opportunities.

Six Sigma methodology is not so commonly used in construction projects, however DMAIC tool can be applied at various stages in construction projects. These are:

1. Detailed Design Stage—To enhance coordination method in order to reduce repetitive work
2. Construction Stage—Preparation of builder's workshop drawings and composite drawings, as it needs a lot of coordination among different trades.
3. Construction Stage—Preparation of contractor's construction schedule.
4. Execution of works.

4.2.9.3.4 DMADDD Tool

Define → Where must we be leaner?
Measure → What's our baseline?
Analyze → Where can we free capacity and Improve yields?
Design → How should we implement?
Digitize → How do we execute?
Draw → Down How do we eliminate parallel paths?

DMADDD TOOL is primarily used to drive the cost out of a process by incorporating Digitization improvements. These improvements can drive efficiency by identifying non-value-added tasks and using simple web-enabled tools to automate certain tasks and improve efficiency. In doing so, employees can be freed up to work on more value-added tasks.

TABLE 4.18 lists the fundamental objectives of DMADDD.

TABLE 4.18
Fundamental Objectives of Six Sigma DMADDD Tool

DMADDD	Phase	Fundamental Objective
1	**Define**—Where must we be leaner?	Identify potential improvements.
2	**Measure**—What's our baseline?	Analog touch points.
3	**Analyze**—Where can we free capacity and improve yields?	Task elimination and consolidated operations. Value-added/non-value-added tasks. Free capacity and yield.
4	**Design**—How should we implement?	Future state vision. Define specific projects. Define draw-down timing. Define commercialization plans.
5	**Digitize**—How do we execute?	Execute project.
6	**Draw Down**—How do we eliminate parallel paths?	Commercialize new process. Eliminate parallel path.

4.2.9.3.4.1 *Impact of Six Sigma Strategy*

The Six Sigma strategy affects five fundamental areas of business. These are:

1. Process improvement.
2. Product and service improvement.
3. Customer satisfaction.
4. Design methodology.
5. Supplier improvement.

4.2.9.4 Application of Six Sigma DMADV Tool for Design of a Building

Six Sigma DMADV tool can be used to develop a design for a building construction project. Following is an example procedure to develop a design for a typical building construction project using Six Sigma DMADV analytic toolset.

DMADV stands for:

1. D → Define
2. M → Measure
3. A → Analyze
4. D → Design
5. V → Verify

DMADV Tool

1. Define → What is Important?
2. Measure → What is needed?
3. Analyze → How will we fulfill?
4. Design → How do we build it?
5. Verify → How do we know it will work?

DMADV Tool has the following usages:

1. DMADV tool is used primarily for the invention and innovation of modified or new products, services, or process.
2. DMADV tool is used when a product, service, or process is required but does not exist.
3. DMADV is proactive, solving problems before they start. This tool is also called DFSS (Design for Six Sigma).
4. Using this tool set Black Belts optimize performance before production begins.

TABLE 4.16 discussed earlier illustrates fundamental objectives of DMADV Tool.

4.2.9.4.1 *Define Phase*

What is important? (Define the project goals and customer deliverables.)

Key deliverables of this phase are:

- Establish the goal.
- Identify the benefits.
- Select project team.
- Develop project plan.
- Develop project charter.

4.2.9.4.1.1 Goal: Develop Construction Project Design Using Six Sigma Tool to Meet Owner's Needs and Satisfy the Project Quality Requirements

4.2.9.4.1.2 Benefits: The Measurable Benefits Adopting This Process Will

- Minimize design errors.
- Minimize omissions.
- Reduce design rework.
- Reduce risk and liabilities.
- Construction project (building) design that will meet owner's needs without rejection of documents by the owner's review team (Project Manager, Construction Manager) at the first submission itself and without affecting the construction process thus reducing external failure cost.
- It will not affect the construction process thus reducing external failure cost.
- Reduce construction overruns.
- Increase customer satisfaction.
- Improve reputation.
- Increase profits.

4.2.9.4.1.3 Selection of Team

The team shall consist of:

1. Sponsor—Project Manager
2. Champion:
    - Design Manager (Civil/Structural)
    - Quality Manager
3. Team Leader—Principal Engineer (Civil/Structural)
4. Team Members—Structural Design Engineer, CAD Technician, Quality Control Engineer, Quantity Surveyor, Cost Engineer, Planner, Owner's Representative, and End User

Similar teams can be organized for other trades such as architectural, HVAC, mechanical (P&FF), electrical including low voltage systems, landscape, and external works.

FIGURE 4.40 illustrates the Design Management Team and their major responsibilities.

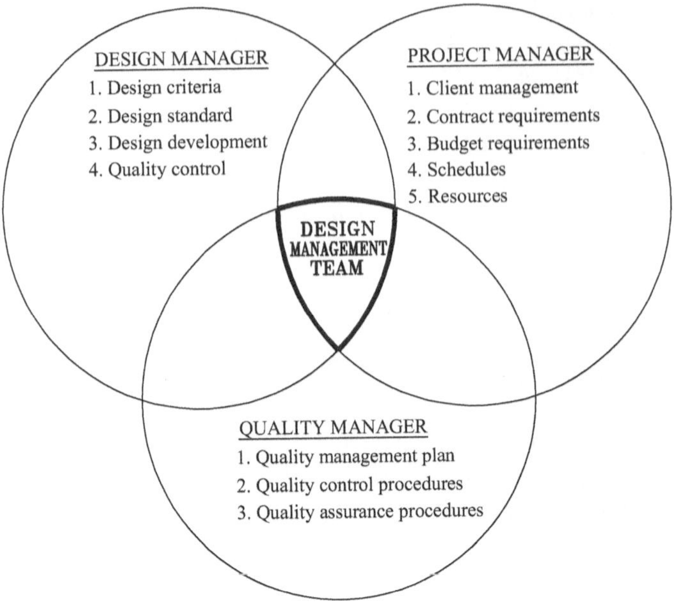

FIGURE 4.40 Design management team.

4.2.9.4.1.4 Project Plan: Time Frame in the Form of Gantt Chart Shall Be Prepared to Meet the Target Dates for

- Data collection and analysis.
- Development of concept design.
- Submission and approval of concept design for client review and comments.
- Development of schematic design.
- Submission and approval of schematic design for client review and comments.
- Development of detailed design (design development).
- Regulatory/authority approvals.
- Preparation of contract documents.
- Submission of detailed design and documents for client review and comments.
- Submission of the project design and documents (final).
- Review and coordination meetings (both internal and with client).
- Monitoring of design schedule.
- Monitoring and controlling resources.

4.2.9.4.1.5 Project Charter: In This Example, the Project Objective Is to Create Construction Project (Building) Design Within Stipulated Schedule That

- Will meet owner's need.
- Will meet relevant standards, codes, regulatory requirements.

- Will be approved by the owner/owner's representative at the first submission without any major comments on the design and specification documents.
- The documents shall be accurate and shall not affect construction quality.

4.2.9.4.2 Measure Phase

What is needed? (Measure and determine customer needs and specifications.)

Key deliverables in this phase are:

1. Matrix of owner's requirements (Terms of Reference (TOR))—Scope of work.
2. Codes and standards to be followed.
3. Regulatory requirements.
4. Sustainability requirements (economic, environmental, and social).
5. Implementation of LEED requirements.
6. Energy conservation requirements.
7. Waste management.
8. Maintainability.
9. Constructability.
10. HSE requirements.
11. Fire protection requirements.
12. Project schedule.
13. Project budget.
14. Design review procedure by the owner/client.
15. Other disciplines' requirements for coordination purposes.
16. Procedure to incorporate changes/revisions requested by the client/owner.
17. Number of drawings to be produced.
18. Duration of completion of project design and submission of documents at different stages/phases of the project.
19. Quality system requirements within the organization or portion of the organization.
20. Any special condition by the client.

4.2.9.4.3 Analyze Phase

How do we fulfill? (Analyze process options and prioritize based on capability to satisfy customer requirements.)

- Key deliverables in this phase are:
- Data collection.
- Prioritization of data under major variables.
- Compliance with the organization's procedures and guidance.

4.2.9.4.3.1 Data Collection

Data shall be collected at different phases of the construction project life cycle, for design considerations, and to meet requirements for the development of the following phases:

1. Concept Design
2. Schematic/Preliminary Design
3. Detailed Design/Design Development

The following data shall be collected to develop construction project design at different phases:

- Identification of need by the owner.
- Identify the number of floors.
- Identify building usage.
- Identify technical and functional capability requirements.
- Soil profile and laboratory test of soil.
- Identify topography of the project site.
- Identify wind load, seismic load, dead load, and live load.
- Identify existing services passing through the project site.
- Identify existing roads, structure surrounding the project site.
- Identify environmental compatibility requirements.
- Identify energy conservation requirements.
- Identify sustainability requirements.
- Identify regulatory/authority requirements.
- Identify codes and standards to be followed.
- Identify social responsibility requirements.
- Identify health and safety features.
- Identify fire protection requirements.
- Identify aesthetic requirements.
- Identify zoning requirements.
- Identify project constraints.
- Identify ease for constructability.
- Identify critical activities during construction.
- Identify method statement requirements.
- Identify requirements of other disciplines for coordination purposes.
- Identification of team members.
- Project time schedule.
- Financial implications and resources for the project.
- Identify cost-effectiveness of the project.
- Identify 3D information areas.
- Identify suitable software programs.
- Identify number of drawings to be produced.
- Identify milestones for the development of each phase of design.

4.2.9.4.3.2 Arrangement of Data

The generated data shall be prioritized in an orderly arrangement under the following major variables during the designing of concept, schematic, and design development phases:

- Owner's need.
- Regulatory compliance.
- Sustainability.
- Safety.
- Constructability.

4.2.9.4.3.4 *Design Phase*

How do we build it? (Design detailed process(es) capable of satisfying customer requirements.)

Key deliverables in this phase are:

- Development of concept design.
- Development of schematic design.
- Development of detailed design.

4.2.9.4.3.4.1 Development of Concept Design

While developing the concept design, the designer shall consider the following:

- Project goals/owner needs.
- Number of floors.
- Usage of building.
- Technical and functional capability.
- Regulatory requirements.
- Authorities' requirements.
- Environmental compatibility.
- Sustainability requirements.
- Constructability.
- Health and safety.
- Cost effectiveness over the entire life cycle of the project.

4.2.9.4.3.4.2 Development of Schematic Design

While developing the schematic design, the designer shall consider the following:

- Concept design deliverables.
- Site location in relation to the existing environment.
- Building structure.
- Floor grade and system.
- Tentative size of columns, beams.
- Stairs.
- Roof.
- Authorities' requirements.
- Energy conservation issues.
- Available resources.
- Environmental issues.
- Sustainability requirements.
- Requirements of all stakeholders.
- Optimized life cycle cost (value engineering).
- Constructability.
- Functional/aesthetic aspect.
- Services requirements.

- Project schedule.
- Project budget.
- Preliminary contract documents (outline specifications).

4.2.9.4.3.4.3 Development of Detailed Design

While developing the detailed design, the designer shall consider the following:

- Schematic design deliverables.
- Authorities' requirements.
- Energy conservation issues.
- Environmental issues.
- Sustainability requirements.
- Requirements of all stakeholders.
- Environmental compatibility.
- Available resources.
- Number of floors
- Property limits/surrounding areas.
- Excavation.
- Dewatering.
- Shoring.
- Backfilling.
- Substructure.
- Design of foundation based on field and laboratory tests of soil investigation.
- Subsurface profiles and subsurface conditions, and subsurface drainage.
- Co-efficient of sliding on the foundation.
- Degree of difficulty for excavation.
- Method of protecting below-grade concrete members against the impact of soil and groundwater.
- Geotechnical design parameters.
- Design load, such as dead load, live load, and seismic load.
- Grade and type of concrete.
- Type of footings.
- Type of foundation.
- Energy efficient foundation.
- Size of bars for reinforcement and the characteristic strength of bars.
- Clear cover for reinforcement.
- Reinforcement bar schedule, stirrup spacing.
- Superstructure.
- Columns.
- Walls.
- Stairs.
- Beams.
- Slab.
- Parapet wall.
- Height of each floor.

- Beam size and height of beams.
- Location of columns in coordination with architectural requirements.
- Openings for services.
- Deflection which may cause fatigue of structural elements, crack or failure of fixtures, fittings, or partitions, or discomfort of occupants.
- Movement and forces due to temperature.
- Equipment vibration criteria.
- Expansion joints.
- Insulation.
- Concrete tanks (water storage).
- Services requirements (shafts, pits).
- Shafts and pits for conveying system.
- Building services to fit in the building.
- Coordination with other trades and conflict resolution.
- Calculations required as per contract requirements.

4.2.9.4.3.4.4 Preparation of Contract Documents

Contract documents shall be prepared as per MasterFormat® (©2022).

4.2.9.4.3.5 Verify Phase

How do we know it will work? (Verify design performance capability.)

Key deliverables in this phase are:

- Review and check the design for quality assurance using thorough itemized review checklists to ensure that design drawings fully meet the owner's objectives/goals.
- Review and check contract documents (design drawings, specifications, and contract documents).
- Check for accuracy.
- Check calculations.
- Review studies and reports.
- Review discipline requirements.
- Review inter-discipline requirements and conflict.
- Review constructability.
- Review for sustainability compliance.
- Management review.

After verification, the documents can be released for submission.

In case of any comments from the client/client's representative, the design shall be reviewed and modified accordingly.

4.2.9.5 Application of Six Sigma DMADV Tool for Construction Schedule

The contractor's construction schedule (CCS) is an important document used during the construction phase. It is used to plan, monitor, and control project activities and resources. The document is voluminous and important. It has to be prepared with accuracy in order to follow the work progress without deviation from the milestones set up in the contract documents. Generally, the project interim payment to the contractor is linked to the approval of the CCS. The contractor is not paid unless the CCS is approved by the construction manager/project manager/consultant.

In most cases, contractors experience problems with getting the CCS approved, at the very first submission, from the construction manager/project manager/consultant. It could be rejected if it does not meet the specifications. Therefore, the contractor has to put all effort into collecting relevant data to be fed to develop the CCS.

4.2.9.5.1 The DMADV Process

The following is an example procedure to develop the CCS using the Six Sigma DMADV analytic tool set. The DMADV method is used primarily for the invention of modified or new products, services, or processes. DMADV stands for:

1. Define → What is important?
2. Measure → What is needed?
3. Analyze → How will we fulfill?
4. Design → How do we build?
5. Verify → How do we know it will work?

4.2.9.5.1.1 Define Phase (What Is Important?)

The objective of this phase is to define the project goals and customer deliverables.

The key deliverables of this phase are:

- Establish the goal.
- Identify the benefits.
- Select project team.
- Develop a project plan.

Goal: Develop CCS using Six Sigma tools.

Benefits: The measurable benefits in adopting this process will result in CCS that will meet all the requirements of the specifications and shall be approved by the construction manager/project manager/consultant at the first submission itself. This will reduce the repetitive work and help implement the schedule right from the early stage of the project.

Selection of team: The team shall consist of

1. Sponsor—Project Manager
2. Champion—Construction Manager
3. Team Leader—Planning and Control Manager
4. Team Members—Planning Engineer, Cost Engineer, and One Representative from Each Subcontractor

Project plan: Time frame in the form of a Gantt chart shall be prepared to meet the target dates for submitting the contractor's construction schedule (CCS).

4.2.9.5.1.2 Measure Phase (What Is Needed?)

The objective of this phase is to measure and determine customer needs and specifications.

The key deliverables in this phase are:

- Identify specification requirements.

The following are the requirements listed in most contract documents.

The contractor has to submit the construction schedule in a bar chart time-scaled format to show the sequence and interdependence of activities required for the complete performance of all items of work under the contract. The contractor shall use a computerized precedence diagram critical path method (CPM) technique in preparation of CCS. The schedule shall include, but not be limited to, the following:

1. Project site layout.
2. Concise description of the work.
    a. Milestones (contractual milestones or constraints).
    b. Number of working days.
    c. Work breakdown structure (WBS) activities shall consist of all those activities that take time to carry out execution/installation and on which resources are expended.
    d. Construction network of project phases (if any), including various subphases.
    e. Construction network of the project arrangements (activities) and sequence.
    f. Time schedules for various activities in a bar chart format.
    g. The minimum work activities to be included in the program shall include items stated in the bill of quantities (BOQ).
    h. WBS activities shall consist of all those activities that take time to carry out execution/installation and on which resources are expended.
    i. Early and late finish dates.
    j. Time schedule for the critical path.
    k. Schedule text report showing activity, start and finish dates, total float, and relationship with other activities.
    l. Summary schedule report showing the number of activities, project start, project finish, number of relations, open ends, constraints, and milestones.
    m. Total float of each activity.
    n. Cost loading.
    o. Expected progress cash flow S-curve.
    p. Resource-loaded S-curve.
    q. Manpower loading.
    r. Labor and crew movement and distribution.
    s. Resource productivity schedule.

t. The number of hours per shift.
u. Average weekly usage of manpower for each trade.
v. Resource histogram showing the manpower required for different trades per time period for each trade (weekly or monthly).
w. Equipment and machinery loading.
x. Schedule of mobilization and general requirements.
y. Schedule of subcontractors and suppliers' submittal and approval.
z. Schedule of materials submittals and approvals.
aa. Schedule of long lead materials.
bb. Schedule of procurement.
cc. Schedule of shop drawings submittals and approvals.
dd. Regulatory/authorities' requirements.
ee. Schedule of testing, commissioning, and handover
ff. Expected cash flow for executed work (during progress of work).

4.2.9.5.1.3 Analyze Phase (How Will We Fulfill?)

The objective of this phase is to analyze process options and prioritize based on the capability to satisfy customer requirements.

The key deliverables in this phase are:

- Data collection.
- Prioritization of data under major variables.

Data collection: The objectives of this process are to:

1. Identify milestone dates and constraints
2. Identify project calendar.

   a. Identify resource calendar.
   b. Review contract conditions and technical specifications.
   c. Identify mobilization requirements.
   d. Identify project method statement.
   e. Identify subcontractors/suppliers.
   f. Identify materials requirements.
   g. Identify long lead items.
   h. Identify procurement schedule.
   i. Identify shop drawing requirements.
   j. Identify regulatory/authorities' requirements.
   k. Identify WBS activities using BOQ (Bill of Quantity).
   l. Relate WBS activities with BOQ and contract drawings.
   m. Identify zoning/phasing.
   n. Identify codes for all activities per contract document divisions/sections per the Construction Specifications Institute (CSI) format.
   o. Identify volume of work for each activity.
   p. Identify duration/time schedule of each activity.

q. Identify early and late finish dates.
r. Identify critical activities and their effect on critical path.
s. Identify logical relationship.
t. Identify sequencing of activities.
u. Identify project progress cash flow (work in place).
v. Identify manpower resources with productivity rate.
w. Identify equipment and machinery.
x. Identify project constraints such as access, logistics, delivery, seasonal, national, safety, existing workflow discontinuity, and proximity of adjacent concurrent work.
y. Identify testing, commissioning, and handover requirements.
z. Identify special inspection requirements.
aa. Identify closeout requirements.
bb. Identify and include items not listed in the specifications but are important for project scheduling.
cc. Identify suitable software programs.
dd. Identify submittal requirements.

Arrangement of data: The generated data can be prioritized in an orderly arrangement under the following major variables:

1. Milestones
2. WBS activities

   a. Time schedule.
   b. General requirements.
   c. Resources.
   d. Engineering.
   e. Cost loading.

FIGURE 4.41 illustrates these variables along with related sub variables arranged in the form of the Ishikawa diagram.

4.2.9.5.1.4 Design Phase (How Do We Build It?)

The objective of this phase is to design detailed processes capable of satisfying customer requirements.

The key deliverables in this phase are:

- Preparation of program using suitable (specified) software programs.

The project and control manager can prepare the CCS based on the collected data and sequence of activities.

4.2.9.5.1.5 Verify Phase (How Do We Know It Will Work?)

The objective of this phase is to verify the design performance capability.

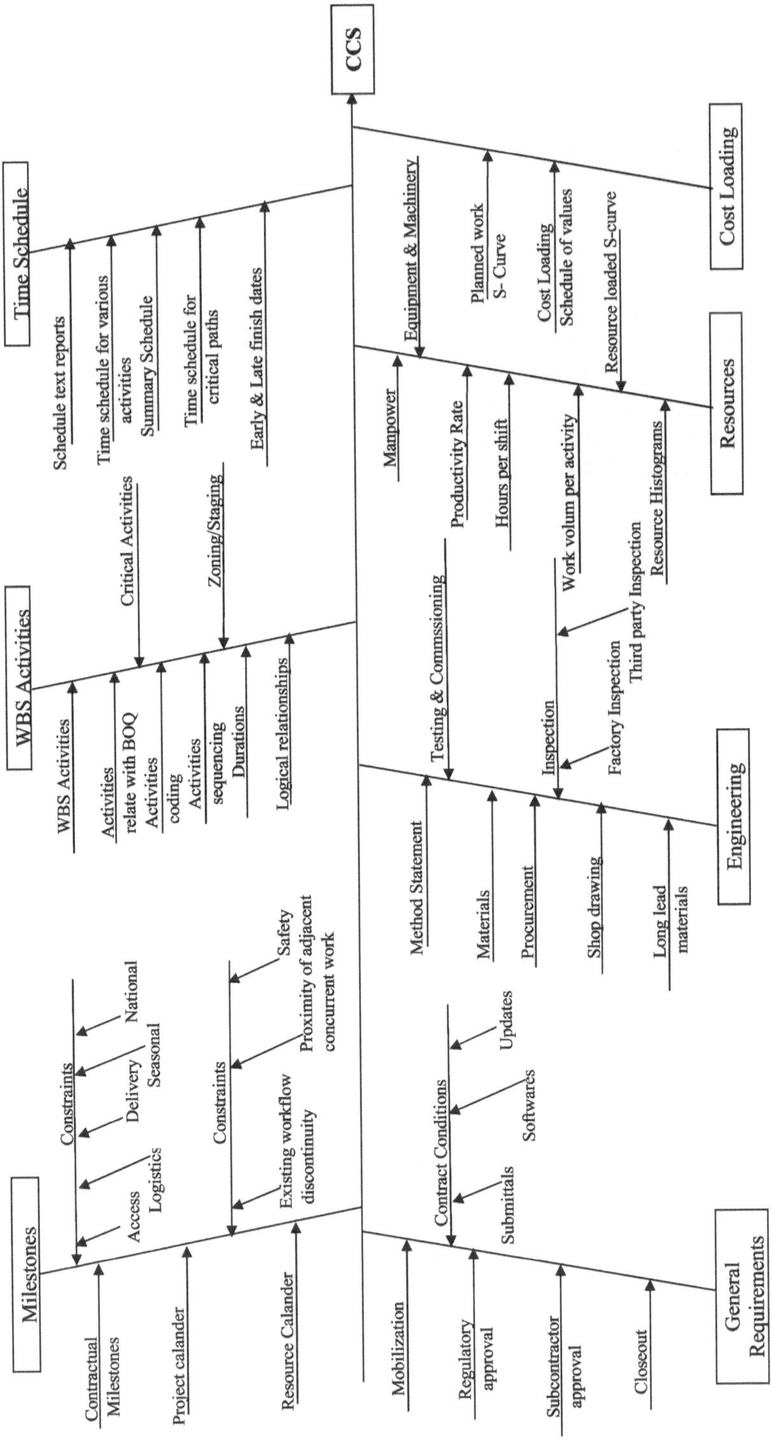

FIGURE 4.41 Ishikawa diagram for CCS data.

The key deliverables in this phase are:

- Review the schedule by the team members to ascertain that all the required elements are included for compliance with specification requirements.
- Submit CCS to construction manager/project manager/consultant.
- Update the schedule as and when required.

4.2.10 TRIZ

TRIZ is short for teirija rezhenijia izobretalenksh zadach (theory of inventive problem solving), developed by the Russian Scientist Genrish Altshuller. TRIZ provides systematic methods and tools for analysis and innovative problem-solving to support decision-making process.

Continuous and effective quality improvement is critical for an organization's growth, sustainability, and competitiveness. The cost of quality is associated with both chronic and sporadic problems. Engineers are required to identify, analyze the causes, and solve these problems by applying various quality improvement tools. Any of these quality tools taken individually do not allow a quality practitioner to carry out the whole problem-solving cycle. These tools are useful for solving a particular phase of the problem and need a combination of various tools and methods to find a problem solution. TRIZ is an approach that starts at a point where fresh thinking is needed to develop a new or redesign a process. It focuses on methods for developing ideas to improve a process, get something done, design a new approach, or redesign an existing approach. TRIZ offers a more systematic although still universal approach to problem solving. TRIZ has advantages over other problem-solving approaches in terms of time efficiency and has low-cost quality improvement solutions. The pillar of TRIZ is the realization that contradictions can be methodically resolved through the application of innovative solutions. Altshuller defined an inventive problem as one containing a contradiction. He defined contradiction as a situation where an attempt to improve one feature of system detracts from another feature.

4.2.10.1 TRIZ Methodology

Traditional processes for increasing creativity have a major flaw in that their usefulness decreases as the complexity of the problem increases. At times, trial and error methods are used in every process and the number of trials increases with the complexity of the inventive problem. In 1946 Altshuller determined to improve the inventive process by developing the "science" of creativity, which led to the creation of TRIZ. TRIZ was developed by Altshuller as a result of analysis of many thousands of patents. He reviewed over 200,000 patents looking for problems and how they are solved. He selected 40,000 as representative of inventive solutions, the rest were direct improvements easily recognized within the system. Altshuller recognized a pattern where some fundamental problems were solved with solutions that were repeatedly used from one patent to another, although the patent subject, applications, and timings varied significantly. Altshuller categorized these patterns into five levels of inventiveness. TABLE 4.19 summarizes Altshuller's findings.

TABLE 4.19
Levels of Incentives

Level	Degree of Inventiveness	% of Solutions	Source of Solution
1	Obvious solution.	32%	Personal skill.
2	Minor improvement.	45%	Knowledge within existing systems.
3	Major improvement.	18%	Knowledge within the industry.
4	New concept.	4%	Knowledge outside industry and found in science, not in technology.
5	Discovery.	1%	Outside the confines of scientific knowledge.

He noted that with each succeeding level, the source of the solution required broader knowledge and more solutions to consider before an ideal solution could be found.

TRIZ is a creative thinking process that provides a highly structured approach for generating innovative ideas and solutions for problem-solving. It provides tools and methods for use in problem formulation, system analysis, failure analysis, and pattern of system evolution. TRIZ is a contrast to techniques such as brainstorming, which aims to create an algorithmic approach to the invention of new systems and refinement of old systems. Using TRIZ requires some training and a good deal of practice.

TRIZ's body of knowledge contains 40 creative principles drawn from analysis of how complex problems have been solved.

- The laws of systems solution.
- The algorithm of inventive problem-solving.
- Substance-field analysis.
- 76 standard solutions.

4.2.10.2 Application of TRIZ

Engineers can apply TRIZ for solving the following problems in construction projects:

- Non-availability of specified material.
- Regulatory changes to use certain types of material.
- Failure of dewatering system.
- Casting of a lower grade of concrete to that of a specified higher grade.
- Collapse of trench during excavation.
- Collapse of formwork.
- Collapse of roof slab while casting in progress.
- Chiller failure during peak hours in the summer.
- Modifying method statement.
- Quality auditors can use TRIZ to develop corrective actions to the audit findings during auditing.

4.2.10.3 TRIZ Process

Altshuller has recommended four steps to invent a new solution to a problem. These are:

Step 1—Identify the problem.
Step 2—Formulate the problem.
Step 3—Search for a precisely well-solved problem.
Step 4—Generate multiple ideas and adapt a solution.

The aforementioned methods are primarily used for low-level problems. To solve more difficult problems more precise tools are used. These are as follows:

1. ARIZ (Algorithm for Inventive Problem Solving)
2. Separation Principles
3. Substance-Field Analysis
4. Anticipator Failure Determination
5. Direct Product Evaluation

Quality Function Deployment (QFD) matrix is also used to identify new functions and performance levels to achieve a truly exciting level of quality by eliminating technical bottlenecks at the conceptual stage. QFD may be used to feed data into TRIZ, especially using the "rooftop" to help develop contradictions.

The different schools for TRIZ and individual practitioners have continued to improve and add to the methodology.

(Source: Abdul Razzak Rumane (2010), *Quality Management in Construction Projects*, reprinted with permission from Taylor & Francis Group).

4.3 TOOLS FOR CONSTRUCTION PROJECT DEVELOPMENT

Project development is a process spanning from the commencement of project initiation and ends with closeout and finalizing project records after project construction and handing over of the project. The project development process is initiated in response to an identified need. It covers a range of time-framed activities extending from identification of a project need to a finished set of contract documents and to construction.

Construction project development has four major elements/stages. These are:

1. Study
2. Design
3. Bidding and Tendering
4. Construction

As the project develops, more information and specifications are developed.

TABLE 4.20 illustrates the construction project development stages.

4.3.1 Tools for Study Stage

TABLE 4.21 illustrates major elements of the study stage and related tasks performed during this stage.

TABLE 4.20
Construction Project Development Stages

Sr. No.	Stages	Elements	Description
1	Study	Problem Statement/Need Identification	Project needs, goals, and objectives.
		Need Assessment	Identification of needs.
			Prioritization of needs.
			Leveling of needs.
			Deciding what needs to be addressed.
		Need Analysis	Perform project need analysis/study to outline the scope of issues to be considered in the planning phase.
		Need Statement	Develop project need statement.
		Feasibility Study	Technical studies, economics assessment, financial assessment, scheduling, market demand, risk, environmental and social assessment.
		Establish Project Goals and Objectives	Scope, time, cost, quality.
		Identify Alternatives	Identify alternatives based on a predetermined set of performance measures.
		Preliminary Schedule	Estimate the duration for completion of project/facility.
		Preliminary Financial Implications	Preliminary budget estimates of total project cost (life cycle cost) on the basis of any known research and development requirements. This will help arrange the finances (funding agency).
		Preliminary Resources	Estimate resources.
		Project Risk	Risk, constraints.
		Authorities' Clearance	Identify issues, sustainability, impacts, and potential approvals (environmental, authorities, permits) required for subsequent design and authority approval processes.
		Select Preferred Alternative	Select preferred alternative considering technological and economical feasibility.

(Continued)

TABLE 4.20 (Continued)
Construction Project Development Stages

Sr. No.	Stages	Elements	Description
		Identify Project Delivery System	Establish how the participants, owner, designer (A/E), and contractor will be involved to construct the project/facility. (design-build-bid, design-build, guaranteed maximum price, CM type, PM type, BOT, turnkey, etc.).
		Identify Type of Contracting System/ Pricing	Select contract pricing systems such as firm fixed price or lump sum, unit price, cost reimbursement (cost plus), reimbursement, target price, time and material, guaranteed maximum price.
		Develop Project Charter	Terms of reference (TOR).
		Identify Project Team	Select designer (A/E) firm if design-bid-build type of contract system is selected. Select other team members based on project delivery system requirements.
		Project Launch	Proceed with concept design.
2	Design	Develop Concept Design	Report, drawings, models, presentation.
		Regulatory Approvals	Obtain regulatory approvals.
		Project Planning	Prepare project plan.
		Schematic Design	Preliminary design, value engineering.
		Design Development	Detailed design.
		Construction Documents	Construction contract documents.
3	Bidding and Tendering	Bid Documents	Tendering documents.
		Selection of Contractor	Advertisement for bidders.

(Continued)

TABLE 4.20 (Continued)
Construction Project Development Stages

Sr. No.	Stages	Elements	Description
			Pre-qualification of contractors.
			Issuing tender documents/ request for proposal.
			Receipt of bid documents.
			Tabulation of proposals.
			Analysis of proposal.
			Selection of contractor.
			Most competitive bidder (low bid, qualification-based).
		Award Contract	Signing of contract.
			Bonds and insurance.
			Notice to proceed.
4	Construction	Construction	Contractor to execute contracted works.
		Monitoring and Control	Monitor and control scope, schedule, cost, quality, risk, procurement of the project.
		Commissioning and Handover	Testing, commissioning, and handover.
		Project Closeout	Close the project.

TABLE 4.21
Major Elements of Study Stage

Serial Number	Elements	Description	Related Task
1	Problem Statement/ Need Identification	Project needs, goals, and objectives.	Strategic objectives, policies, and priorities (please refer to FIGURE 4.42).
2	Need Assessment	Identification of needs. Prioritization of needs. Leveling of needs. Deciding what needs to be addressed.	Ensure that the owner's business case has been properly considered (please refer TABLE 4.22).
3	Need Analysis	Perform project need analysis/ study to outline the scope of issues to be considered in the planning phase.	Perform need analysis (please refer to TABLE 3.2).
4	Need Statement	Develop project need statement.	Develop need statement (please refer to TABLE 3.3).
5	Feasibility Study	Technical studies, economics assessment, financial assessment, scheduling, market demand, risk, environmental and social assessment.	Perform feasibility study statement (please refer to TABLE 4.23).
6	Establish Project Goals and Objectives	Scope, time, cost, quality.	Project initiation documents developed on SMART concept.
7	Identify Alternatives	Identify alternatives based on a predetermined set of performance measures.	Select conceptual alternatives.
8	Preliminary Schedule	Estimate the duration for completion of project/facility.	Establish project schedule.
9	Preliminary Financial Implications	Preliminary budget estimates of total project cost (life cycle cost) on the basis of any known research and development requirements. This will help arrange the finances (funding agency).	Determine project budget.
10	Preliminary Resources	Estimate resources.	Confirm availability of resources, manpower, material, equipment.
11	Project Risk	Identify project risks, constraints.	Establish risk response, mitigation plan.

(*Continued*)

TABLE 4.21 (Continued)
Major Elements of Study Stage

Serial Number	Elements	Description	Related Task
12	Authorities' Clearance	Identify issues, sustainability, impacts, and potential approvals (environmental, authorities, permits) required for subsequent design and authority approval processes.	Establish requirements for statutory approvals and other regulatory authorities.
13	Select Preferred Alternative	Assess technological and economic feasibility and compare to the preferred option/alternative to prepare business case.	Discuss relative merits of various alternative schemes and evaluate the performance measures to meet owner's needs/ requirements. Consider social, economic, environmental impact, safety, reliability, and functional capability.
14	Identify Project Delivery System	Establish how the participants, owner, designer (A/E), and contractor will be involved in constructing the project/ facility. (design-build-bid, design-build, guaranteed maximum price, CM type, PM type, BOT, turnkey, etc.).	Select suitable project delivery system as per strategic decision and suitability of appropriate system (please refer to Chapter 2).
15	Identify Type of Contracting System	Select contract pricing systems such as firm fixed price or lump sum, unit price, cost reimbursement (cost plus), reimbursement, target price, time and material, guaranteed maximum price.	Select contracting/pricing most appropriate for the benefit of owner (please refer to Chapter 2).
16	Identify Project Team	Select designer (A/E) firm if design-bid-build type of contract system is selected. Select other team members based on project delivery system requirements.	Select project team considering the selected type of project delivery system and procurement process of the organization (please refer to Chapter 2).
17	Project Launch	Project charter.	Prepare terms of reference (TOR) (please refer to TABLE 4.24).

FIGURE 4.42 illustrates the logic flow diagram in the study stage (conceptual design stage).

TABLE 4.22 illustrates points to be considered for need assessment.

TABLE 4.23 illustrates points to be considered for the feasibility study of a construction project.

TABLE 4.24 illustrates the contents of the terms of reference (TOR).

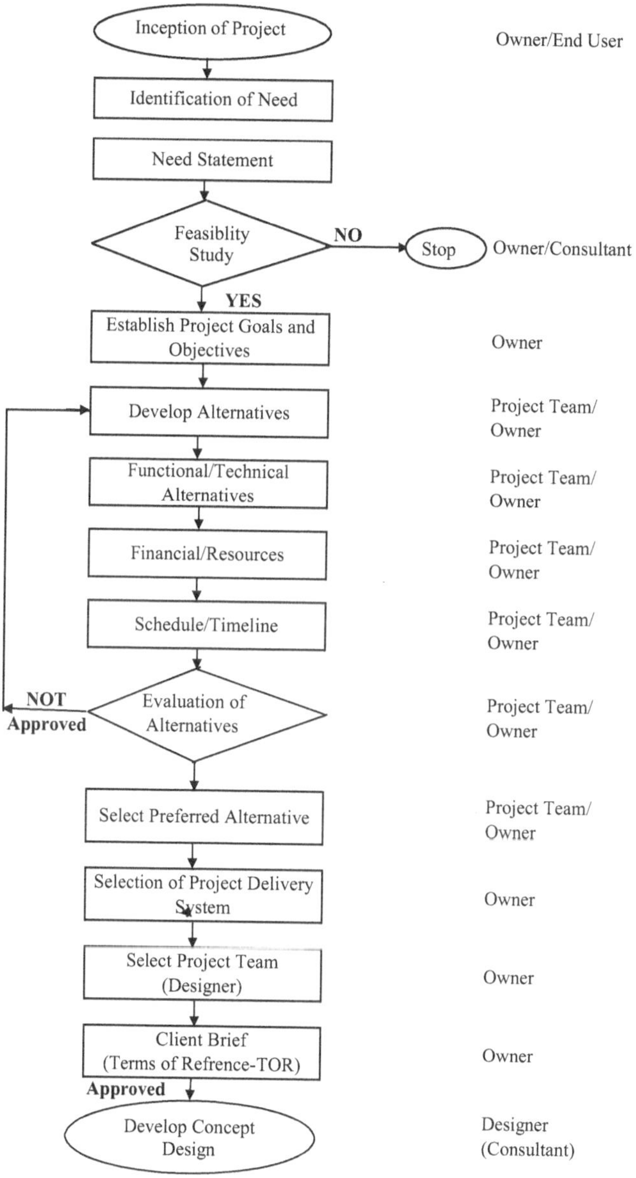

FIGURE 4.42 Logic flow of activities in the study stage.

TABLE 4.22
Need Assessment for Construction Project

Serial Number	Points to be Considered
1	Ensure the need is properly defined.
2	Confirm that the needed outcome will benefit the owner/end user.
3	Gather and analyze owner/end user requirements.
4	Set priorities and establish criteria for solutions to meet the targeted demand or requirements.
5	Identify and measure areas for improvement to achieve the objectives of the project.

TABLE 4.23
Major Considerations for Feasibility Study of a Construction Project

Serial Number	Points to be Considered
1	Technical suitability of facility for intended use by the owner/end user.
2	Economic feasibility to ascertain value of benefit that results from the project exceeds the cost that results from the project.
3	Financial payback period.
4	Market demand.
5	Environmental impact.
6	Social and cultural assessment.
7	Legal and regulatory impacts.
8	Political aspects.
9	Resources availability.
10	Scheduling of the project.
11	Operational.
12	Risk analysis.

TABLE 4.24
Contents of Terms of Reference (TOR) Documents

Serial Number	Topics
1	Project Objectives
	1.1 Background
	1.2 Project Information
	1.3 General Requirements
	1.4 Special Considerations
2	Project Requirements
	2.1 Scope of work
	2.2 Work Program
	2.2.1 Study Phase
	2.2.2 Design Phase
	2.2.3 Tender Stage
	2.2.4 Construction Phase

(*Continued*)

TABLE 4.24 (Continued)
Contents of Terms of Reference (TOR) Documents

Serial Number	Topics
	2.3 Reports and Presentations
	2.4 Schedule of Requirements
	2.5 Drawings
	2.6 Energy Conservation Considerations
	2.7 Cost Estimates
	2.8 Time Program
	2.9 Interior finishes
	2.10 Aesthetics
	2.11 Mechanical
	2.12 HVAC
	2.13 Lighting
	2.14 Engineering Systems
3	Opportunities and Constraints
	3.1 Site Location
	3.2 Site Conditions
	3.3 Land Size and Access
	3,2 Climate
	3.3 Time
	3.4 Budget
4	Performance Target
	4.1 Financial Performance
	4.1.1 Performance Bond
	4.1.2 Insurance
	4.1.3 Delay Penalty
	4.2 Energy Performance Target
	4.2.1 Energy Conservation
	4.3 Work Program Schedule
5	Environmental Considerations
6	Design Approach
	6.1 Procurement Strategy
	6.2 Design Parameters
	6.2.1 Architectural Design
	6.2.2 Structural Design
	6.2.3 Mechanical Design
	6.2.4 HVAC Design
	6.2.5 Electrical Design
	6.2.6 Information and Communication Technology
	6.2.7 Conveying System
	6.2.8 Landscape
	6.2.9 External Works
	6.2.10 Parking
	6.3 Sustainable Architecture
	6.4 Engineering Systems
	6.5 Value Engineering Study
	6.6 Design Review by Client
	6.7 Selection of Products/Systems

(Continued)

TABLE 4.24 (Continued)
Contents of Terms of Reference (TOR) Documents

Serial Number	Topics
7	Specifications and Contract Documents
8	Project Control Guidelines
9	Submittals
	9.1 Reports
	9.2 Drawings
	9.3 Specifications
	9.4 Models
	9.5 Sample Boards
	9.6 Mockup
10	Presentation
11	Project Team Members
	11.1 Number of Project Personnel
	11.2 Staff Qualification
	11.3 Selection of Specialists
12	Visits

TABLE 4.25
5W2H Analysis for Project Need

Serial Number	Why	Related Analyzing Question
1	Why	Why a new project?
2	What	What advantage it will have over other similar projects?
3	Who	Who will be the customer for this project?
4	Where	Where can we find the market for the project?
5	When	When the project will be ready?
6	How many	How many such projects are in the market?
7	How much	How much market share we will have with this project?

Some of the tools discussed earlier can be applied during the study stage. TABLE 4.25 is an example of how 5W2H (under Innovation and Creative Tools) can be used to analyze the project needs.

4.3.2 Tools for Design Stage

TABLE 4.26 illustrates major elements of the design stage and related tasks performed during this stage.

TABLE 4.27 illustrates points to be considered for data collection to facilitate and develop concept design.

TABLE 4.28 illustrates points to be considered for the development of concept design.

FIGURE 4.43 illustrates the preliminary project plan.

TABLE 4.26
Major Elements of Design Stage

Serial Number	Elements	Description	Related Task
1	Develop Concept Design	Report, drawings, models, presentation.	Data collection (please refer to TABLE 4.27). Design development points (please refer to TABLE 4.28).
2	Regulatory Approvals	Obtain regulatory approvals.	Submission of drawings and related documents to authorities and obtain their approvals.
3	Project Planning	Prepare project plan.	Develop project plan considering preliminary schedule for the project (please refer to FIGURE 4.43).
4	Schematic Design	Preliminary design, value engineering.	Develop schematic design (please refer to TABLE 4.29). Preliminary (outline) specifications, authority approvals, value engineering.
5	Design Development	Detailed design.	Develop detailed design (please refer to FIGURE 4.44).
6	Construction Documents	Construction contract documents.	Preparation of particular specifications, contract drawings (please refer to Chapter 6 for details).

TABLE 4.27
Major Items for Data Collection during Concept Design

Serial Number	Items to be Considered
1	Certificate of title: a. Site legalization. b. Historical records.
2	Topographical survey: a. Location plan. b. Site visits. c. Site coordinates. d. Photographs.
3	Geotechnical investigations.
4	Field and laboratory test of soil and soil profile.
5	Existing structures in/under the project site.
6	Existing utilities/services passing through the project site.

(Continued)

TABLE 4.27 (Continued)
Contents of Terms of Reference (TOR) Documents

Serial Number	Items to be Considered
7	Existing roads, structures surrounding the project site.
8	Shoring and underpinning requirements with respect to adjacent area/ structure.
9	Requirements to protect neighboring area/facility.
10	Environmental studies.
11	Daylighting requirements.
12	Wind load, seismic load, dead load, and live load.
13	Site access/traffic studies.
14	Applicable codes, standards, and regulatory requirements.
15	Usage and space program.
16	Design protocol.
17	Scope of work/client requirements.

TABLE 4.28
Development of Concept Design

Serial Number	Points to be Considered
1	Project goals.
2	Usage.
3	Incorporate requirements from collected data.
4	Technical and functional capability.
5	Aesthetics.
6	Constructability.
7	Sustainability (environmental, social, and economic).
8	Health and safety.
9	Reliability.
10	Environmental compatibility.
11	Sustainability.
12	Fire protection measures.
13	Supportability during maintenance/ maintainability.
14	Cost-Effective over the entire life cycle (economy).
15	LEED (equivalent) compliance.
16	Reports, drawings, models.

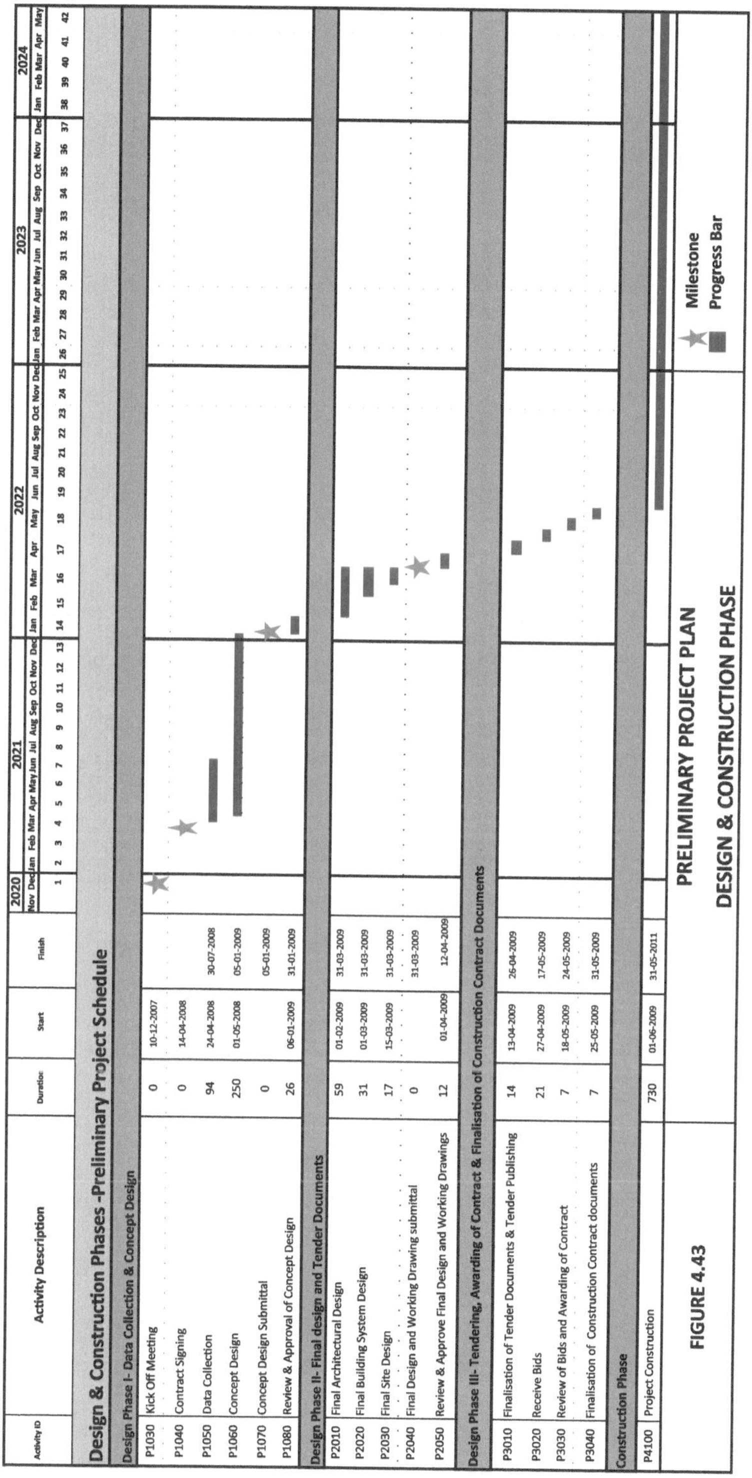

FIGURE 4.43 Preliminary schedule for construction project.

TABLE 4.29
Development of Schematic Design for a Construction Project

Serial Number	Points to be Considered
1	Concept design deliverables.
2	Calculations to support the design.
3	System schematics for electromechanical system.
4	Coordination with other members of the project team.
5	Authorities' requirements.
6	Availability of resources.
7	Constructability.
8	Health and safety.
9	Reliability.
10	Energy conservation issues.
11	Environmental issues.
12	Selection of systems and products that support functional goals of the entire facility.
13	Sustainability.
14	Requirements of all stakeholders.
15	Optimized life cycle cost (value engineering).

TABLE 4.29 illustrates points to be considered for the development of schematic design.

FIGURE 4.44 illustrates major activities in the detailed design phase.

Some of the tools discussed earlier can be applied during the design stage. These are:

- Innovative and Creative Tool
  - Six Sigma DMADV (Please refer to TABLE 4.15)
  - Process Analysis Tool
  - Cost of Quality

- Process Improvement Tool
  - PDCA Cycle
  - Lean Tool
  - Mistake Proofing (Please refer to TABLE 4.14.)

4.3.3 Tools for Bidding and Tendering Stage

TABLE 4.31 illustrates major activities during the bidding and tendering stage and related tasks performed during this stage.

4.3.4 Tools for Construction Stage

TABLE 4.32 illustrates major activities during the construction stage and related tasks performed during this stage.

TABLE 4.33 illustrates the major activities to be performed by the contractor during the construction phase.

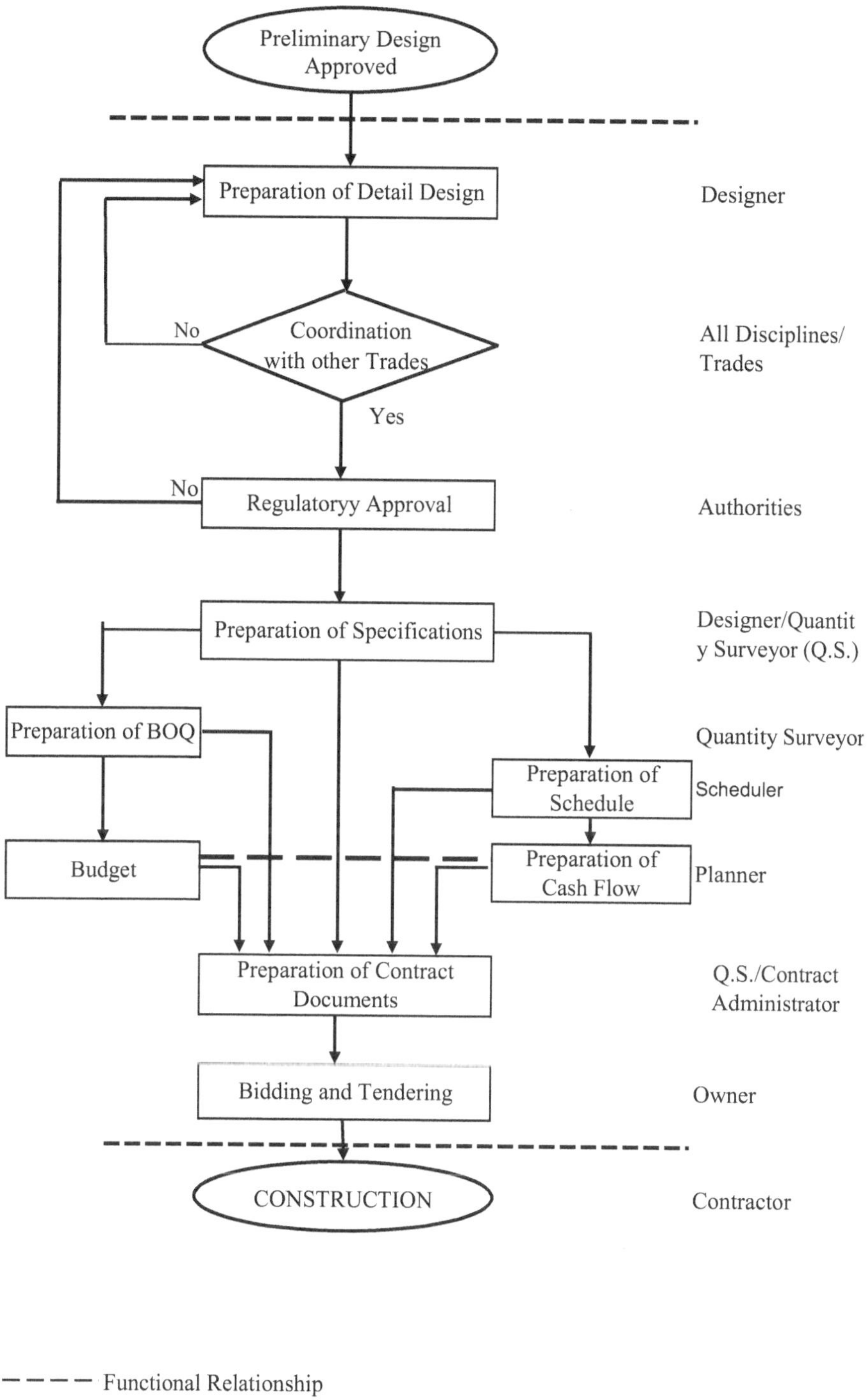

FIGURE 4.44 Major activities in the detailed design phase.

TABLE 4.30
Cost of Quality during Design Stage

Internal Cost	External Cost
• Redesign/redraw to meet fully coordinated design. • Rewrite specifications/documents to meet requirements of all other trades.	• Incorporate design review comments by client/project manager. • Incorporate specifications/documents review comments by client/project manager. • Incorporate comments by regulatory authority(ies). • Resolve RFI (request for information) during construction.
Appraisal Cost	**Prevention Cost**
• Review of design drawings. • Review of specifications. • Review of contract documents to ensure meeting owner's needs, quality standards, constructability, and functionality. • Review for regulatory requirements, codes.	• Conduct technical meetings for proper coordination. • Follow quality system. • Meeting submission schedule. • Training of project team members. • Update of software used for design.

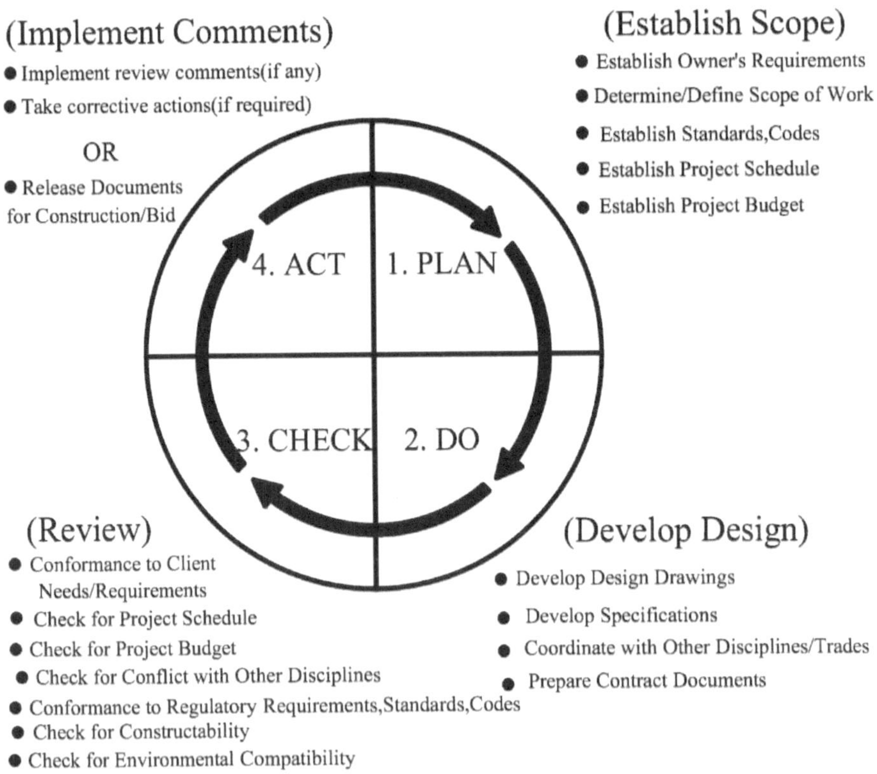

FIGURE 4.45 PDCA cycle for construction project (design stage).

TABLE 4.31
Major Elements of Bidding and Tendering Stage

Serial Number	Elements	Description	Related Task
1	Bidding	Bid Documents	Prepare tender documents.
2	Selection of Contractor	Tender Announcement	Advertisement for bidders. Prequalification of contractors.
3	Issue Tender Documents	Request for Proposal	Issuing tender documents/ request for proposal.
4	Receipt of Proposal	Receipt of Bid Documents	
5	Tender Analysis	Tabulation of Proposals	Analysis of quotation.
6	Selection of Contractor		Low bid. Qualification based. Most competitive bidder (low bid or qualification based as per the company's methodology).
7	Award of Contract	Signing of Contract	Performance bid. Insurance.
8	Notice to Proceed		Contractor to commission.

TABLE 4.32
Major Elements of Construction Stage

Serial Number	Elements	Description	Related Task
1	Construction	Contractor to execute contracted works.	Mobilization, managing resources, work execution, QA/QC, work approval (please refer to TABLE 4.33).
2	Monitoring and Control	Monitor and control scope, schedule, cost, quality, risk, procurement of the project.	Scheduling, monitoring, and control (please refer to FIGURE 4.46).
3	Commissioning and Handover	Testing, commissioning, and handover.	Project startup procedures.
4	Project Closeout	Close the project.	Project closeout documents, finalization of claims and payments.

TABLE 4.33
Major Activities by Contractor during Construction Phase

Serial Number	Activities to be performed by Contractor
1	Mobilization
2	Staff Approval
3	Selection of Subcontractor(s)
4	Selection of Material
5	Selection of Resources
6	Preparation of Shop Drawings
7	Execution of Works
8	Project Monitoring and Control
9	Auditing/Installation of Executed Works
10	Quality Management
11	Approval of Works
12	Health, Safety, and Environmental Compliance

FIGURE 4.46 illustrates the project monitoring and controlling process cycle.

Some of the tools discussed earlier can be applied during the construction stage. These are:

- Management and Planning Tool
  - Networking Arrow Diagram (Please refer to FIGURE 4.12)
  - Process Analysis Tool
  - Root Cause Analysis (FIGURE 4.47)

- Process Improvement Tool
  - Six DMAIC (Please refer to TABLE 4.16)
  - PDCA Cycle (FIGURE 4.48)

- Classic Quality Tool
  - Flow chart for concrete casting (FIGURE 4.4).

4.4 QUALITY INFORMATION SYSTEM

Quality Information System (QIS) is an organized method of collecting, analyzing, storing, and reporting quality-related information for appropriate decision-making.

FIGURE 4.50 illustrates QIS process.

Quality Information System (QIS) is developed based on the organizational requirements to align with goals and objectives to achieve project quality. TABLE 4.34 illustrates a summary of questions and major responses evolved by the researchers about QIS development.

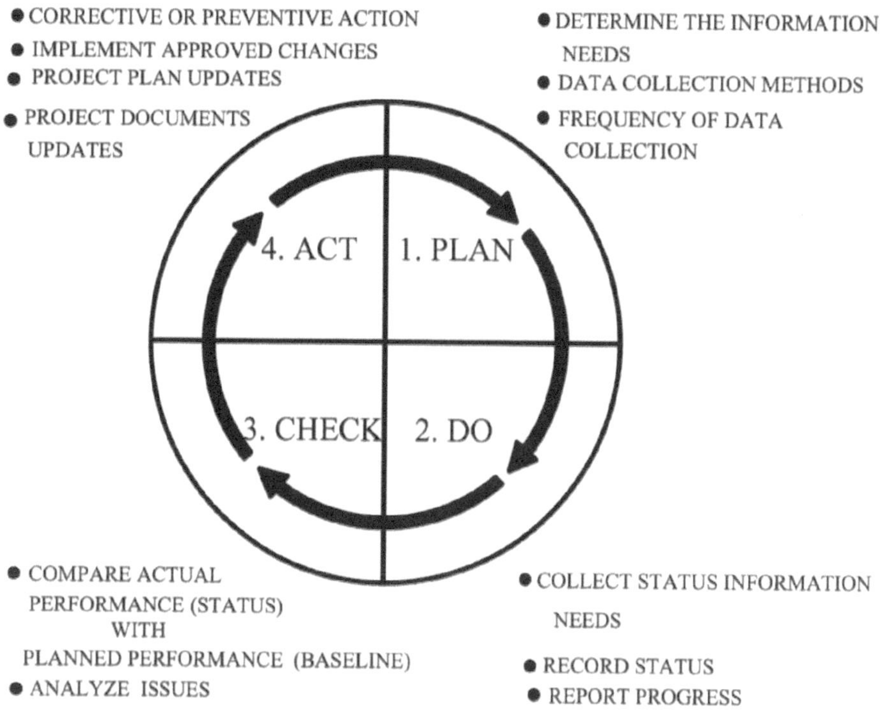

FIGURE 4.46 Project monitoring and controlling process cycle.

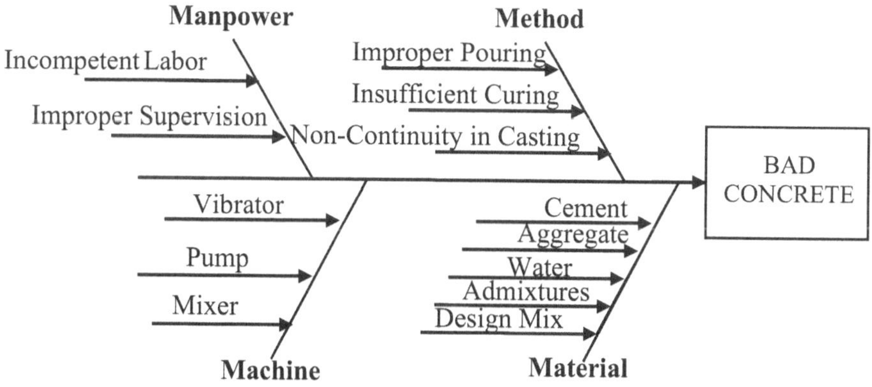

FIGURE 4.47 Root cause analysis for bad concrete.

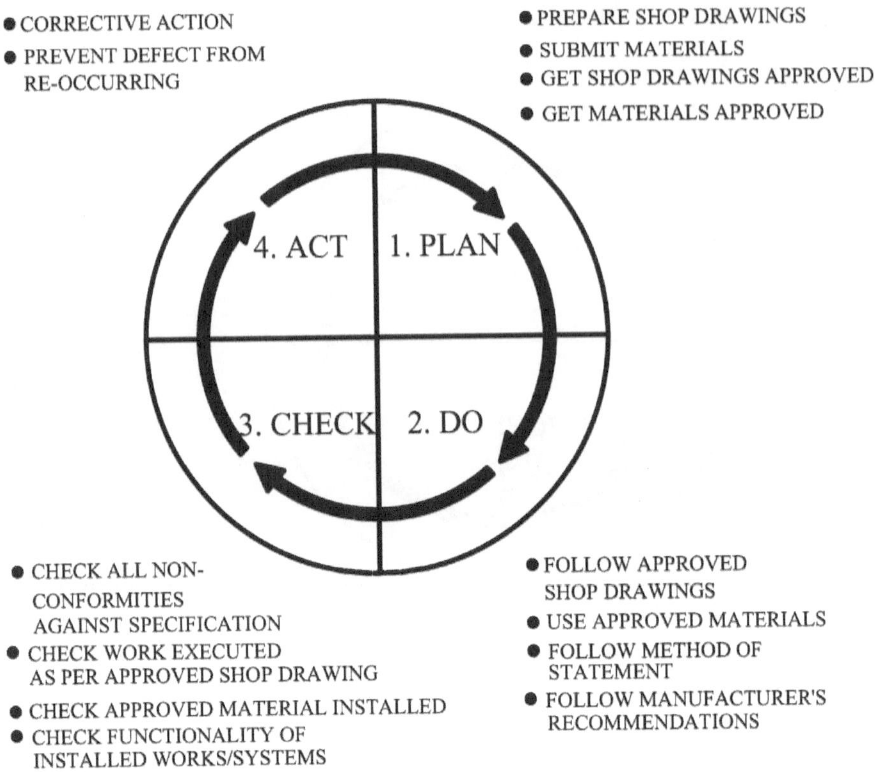

FIGURE 4.48 PDCA cycle (Deming Wheel) for execution of works.

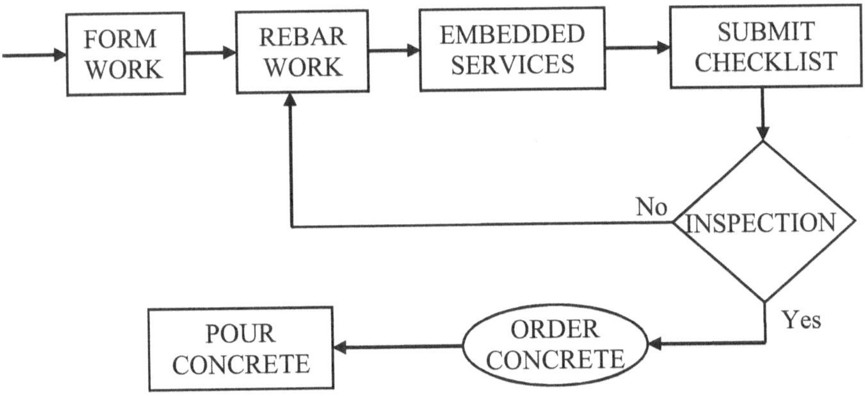

FIGURE 4.49 Flowchart for concrete casting.

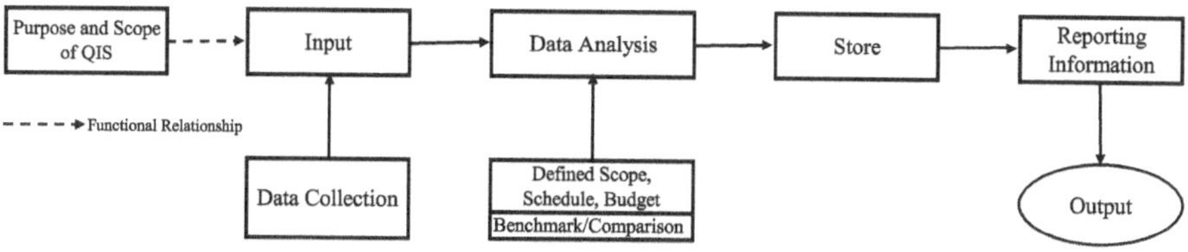

FIGURE 4.50 Quality information system process.

TABLE 4.34
Questions and Major Response for Development of QIS

Sr. No.	Questions	Majority Response
1	Who was the compelling force?	Top-level company officer.
2	Were purposes defined?	Number.
3	How was the information selected?	Issues selected by the user.
4	How was change built into the system?	Did not address change initially. Later incorporated change.
5	Was the system externally benchmarked?	Yes. Companies were compared to external standards.
6	What were the effective ways to analyze performance results?	Comparative charts and training.
7	What were the effective ways to transmit performance results?	Meetings and the Internet with the use of peer pressure.
8	What is the quality of measurements?	All companies used filters to improve usefulness and believability.
9	What measurements are made?	Turnaround time; report length; usefulness.
10	How can the QIS be more effective?	Meeting to share information and research reports from their industry.
11	Was training provided?	Training was provided both during the development process and after.
12	What did not work in thc development process?	Defining the scope of the QIS to be too expansive.
13	What are the most important problems?	1. Relying on the IT department. 2. Lack of ownership of system by employees. 3. No development process flow model.
14	What are the three most important steps to take?	1. Recognition of project importance. 2. Accurate definition of inputs and outputs. 3. Use of project management tools.
15	Can quality be improved in your current system?	Use more technology such as computers and networks to distribute information.

QIS can be manual or computerized. The following section discusses the quality of use of CAD Software.

4.4.1 Quality in Use of CAD Software

With the advent of the computer age, information automation has become a reality. Computers are now used in design, presentation of documents, estimation, presentation, analysis, planning, and many other applications.

Engineering design is the partial realization of the designer's concept. Engineering drawing is an abstract universal language used to represent the designer's concept to others. The conventional method of representing an engineering drawing is drafting on paper with pen or pencil. Manual drafting is tedious and requires a tremendous amount of patience and time. The invention of computer-aided design (CAD) has had a tremendous effect on the design process, and the computer-aided drafting system has improved drafting efficiency. CAD is a process by which a computer-based program assists in the creation or modification of a design. It allows, not only, to pass information more quickly and accurately but also opens the door to integration and automation. A product of the computer era, CAD originated from early comport graphics systems and their development into interactive computer graphics. The 1970s marked the beginning of a new era in CAD—the invention of three-dimensional solid modeling. Since the invention of computer-aided drafting, there has been tremendous progress in CAD, and it is being regularly used in design drafting. In construction projects, CAD is used by all trades such as architectural, structural, mechanical, electrical, infrastructure, and landscape engineers, and others. Design software has helped engineers to produce and modify drawings in the related field of applications and has made it possible to integrate and visualize their overall effects and put them to use to construct the facility. Its application has helped in producing fully coordinated design drawings and avoiding any conflict during construction.

A CAD system consists of three major parts:

1. Hardware: Computer and input/output (I/O) devices.
2. Operating system: Software.
3. Application software: CAD package.

Hardware is used to support the software function. Operating system software is the interface between the CAD application software and the hardware. It is important not only for CAD software but for non-CAD software as well. Application software is the heart of the CAD system. CAD consists of programs that do drafting, engineering analysis, 2-D and 3-D drawings, 3-D modeling, and many other engineering-related functions.

AutoCAD is the most commonly used software in the construction project/industry. It is the most commonly used software by the construction project professional. It began as a PC-based drafting package running under the MS-DOS environment and gradually evolved into a full-blown CAD system. Its editions are upgraded to meet the current demands of industries. AutoCAD 2026 is the latest version of AutoCAD at the time of writing this book.

There are numerous off-the-shelf application packages available; it is beyond the scope of this book to list them all, but they are being used in designing systems in construction projects. The following are the most common application software packages used by all the trades:

- General purpose by all trades:
  - AutoCAD
  - Microsoft Office
- Planning, scheduling, and controlling:
  - Primavera
  - Microsoft Project
- Structural design
  - STADD
  - ETABS
- Mechanical design:
  - ELITE
  - Storm CAD, Water CAD, Sewer CAD
- HVAC design:
  - HAP
  - APEC
  - HEVACOMP
- Electrical design:
  - DIALUX
  - ECODIAL
- Landscape:
  - Dynascape
- Infrastructure/roads:
  - Bently MX
  - Eagle Point

Quality in use aims at defining the quality attributes that are important to the end user. In 1991, the International Organization for Standardization (ISO) introduced ISO/IEC 9126 (1991): Soft-ware evaluation quality characteristics and guidelines for their use. ISO/IEC 9126 is a four-part model:

Part 1: Quality model
Part 2: External metrics
Part 3: Internal metrics
Part 4: Quality in-use metrics

There are six characteristics for both external and internal metrics: functionality, reliability, usability, efficiency, maintainability, and portability. They can be further subdivided into sub-characteristics. The quality in-use metrics has four characteristics: effectiveness, satisfaction, productivity, and safety.

ISO/IEC 9126 was superseded in 2005 by ISO/IEC 25000: Software Engineering. All the requirements of ISO/IEC 9126 have been taken care of in this standard.

TABLE 4.35 lists software quality factors to be considered while selecting a software package.

TABLE 4.35
Software Quality Factors

Sr. No.	Factor	Description
1	Suitability	Whether the application software is suitable and satisfies the intended use and has all the required parameters?
2	Compatibility with the Operating System	Application software is usually operating system dependent, therefore attention must be given to the compatibility of the operating system.
3	Hardware Capability	Whether the RAM and storage memory is sufficient to store the created/modified design data?
4	Reliability	Whether the application software shall perform its intended use with the required precision?
5	Acceptability	Whether the designs produced with application software are accepted by the client?
6	Credibility	How credible the product is?
7	Usability	How easy is it to learn, operate, and transfer input information by translating or interpreting the same to the specific data format?
8	Integrability	Whether the program can be integrated with other application software and how much effort is required for interoperating with other systems?
9	Flexibility	Efforts required to modify an operational system.
10	Efficiency	Whether the program can perform the amount of computing resources and code required by a program?
11	Maintainability	How much effort is required for troubleshooting and fixing an error?
12	Testability	How much effort is required to test and ensure that the program is performing its intended function?
13	Portability	Effort required to transfer the program from one hardware to another and to configure the same with the new environment.
14	Reusability	Whether the program can be used in other applications related to the packaging and scope of the functions that the program performs?
15	Safety	How safe is the program to use?

4.5 QUALITY ASSESSMENT/MEASUREMENT

A quality assessment system in construction projects is a system or method to measure and evaluate the quality of workmanship, finishes, and performance of the constructed work based on approved standards and quality requirements to the satisfaction of the client/end user. In order to achieve competitive advantage and the organization's growth, it is essential the organization has a well-established quality system and effective quality improvement procedure. Quality assessment/measurement needs to be continuous, ongoing, and performed in a timely manner. In construction projects, assessment /measurement is carried out by using checklists, project monitoring, and controlling systems to ensure meeting the the owner needs and is mainly performed during the following phases:

1. Design phase.
2. Construction phase.
3. Testing, commissioning, and handover phase.

Quality assessment/measurement system in construction projects can be classified as:

1. Internal Quality Assessment (Internal Audit)
2. External Quality Assessment

4.5.1 Internal Quality System

The internal quality system is a process in which a quality audit is performed by the organization.

4.5.1.1 Quality Assessment by Designer

Quality assessment during the design phase is normally carried out by the design team as per the quality management system before it is sent for review by the owner (project manager). The designer (consultant) also submits the design to regulatory authorities to obtain their approval for compliance with regulatory requirements prior to the submission of drawings (documents) to the owner/client.

4.5.1.2 Quality Assessment by Contractor

The contractor has to check the executed work prior to submission of the checklist to the supervision engineer (consultant).

4.5.2 External Quality System

External assessment in construction projects can be divided into:

1. Quality inspection/checking by owner-appointed supervision engineer (consultant).

2. Independent testing agency appointed by the owner (consultant as per contract requirements) to inspect/test mainly the following items/works:

   a. Concrete strength testing.
   b. Reinforcement testing.
   c. Elevator work.
   d Chiller testing.
   e. Diesel generator testing.
   f. Electrical switch boards.

The following are the main four steps needed to establish a quality assessment/measurement process in construction projects:

1. Identify the components/items/products/systems to be reviewed and checked in each of the trades (architectural, structural, mechanical, HVAC, electrical, low voltage, landscape, external, etc.).
2. Identify the frequency of checking (timing).
3. Identify the persons/agency authorized to check and approve.
4. Identify the indicators/criteria for performance monitoring.

TABLE 4.36 illustrates a checklist for design drawings. It lists items to be verified and checked internally by the designer before submission to regulatory authorities and subsequently submission to the owner (project manager).

TABLE 4.37 lists major activities to be assessed/checked/inspected by the contractor and supervising engineer to ensure compliance of executed work with contract documents.

TABLE 4.36
Checklist for Design Drawings

Serial Number	Items to be Checked
1	Whether design meets owner requirements (TOR)?
2	Whether design was prepared using authenticated and approved software?
3	Whether design calculation sheets are included in the set of documents?
4	Whether design is fully coordinated for conflict between different trades?
5	Whether design has taken into consideration relevant collected data requirements?
6	Whether reviewer's comments responded to?
7	Whether regulatory approval obtained and comments, if any, incorporated and all review comments responded?
8	Whether design has environmental compatibility?
9	Whether energy efficiency measures are considered?
10	Whether design constructability is considered?
11	Whether design matches with property limits?
12	Whether legends match the layout?
13	Whether design drawings are properly numbered?
14	Whether design drawings have owner logo or designer logo as per standard format?
15	Whether the design format of different trades have uniformity?
16	Whether project name and contract reference are shown on the drawing?

TABLE 4.37
Checklist for Executed Works

Serial Number	Description of Works	Approved	Not Approved	Comments
Shoring				
1	Shoring			
2	Dewatering			
Earth Works				
1	Excavation			
2	Backfilling			
Concrete Sub Structure				
1	Blinding Concrete			
2	Termite Control			
3	Reinforcement of Steel			
4	Concrete Casting			
5	Shuttering for Beams and Columns			
Concrete Super Structure				
1	Reinforcement of Steel for Beams and Columns			
2	Shuttering for Columns and Beams			
3	Concrete Casting of Beams and Columns			
4	Formwork for Slab			
5	Reinforcement for Slab			
6	Concrete Casting of Slab			
7	Precast Panels			
Masonry				
1	Block Work			
2	Concrete Unit Masonry			
Partitioning				
1	Installation of Frames			
2	Installation of Panels			
Metal Work				
1	Structural Steel Work			
2	Installation of Cat Ladders			
3	Installation of Balustrade			
4	Installation of Space Frame			

(*Continued*)

TABLE 4.37 (Continued)
Checklist for Executed Works

Serial Number	Description of Works	Approved	Not Approved	Comments
5	Installation of Handrails and Railings			
Thermal and Moisture Protection				
1	Applying Waterproofing Membrane			
2	Installation of Insulation			
Doors and Windows				
1	Aluminium Windows and Doors			
2	Glazing			
3	Steel Doors and Doors			
4	Wooden Frames			
5	Wooden Doors			
6	Curtain Wall			
7	Hatch Doors			
8	Louvers			
Internal Finishes				
1	Plastering			
2	Painting			
3	Cladding—Ceramic			
4	Cladding—Marble			
5	Ceramic Tiling			
6	Stone Flooring			
7	Acoustic Ceiling			
8	Demountable Partitions			
Furnishing				
1	Carpeting			
2	Blinds			
3	Furniture			
External Finishes				
1	Painting			
2	Brickwork			
3	Stone			
4	Cladding—Aluminium			

(Continued)

TABLE 4.37 (Continued)
Checklist for Executed Works

Serial Number	Description of Works	Approved	Not Approved	Comments
5	Cladding—Granite			
6	Curtain Wall			
7	Glazing			
Equipment				
1	Installation of Maintenance Equipment			
2	Installation of Kitchen Equipment			
3	Installation of Parking Control			
Roof				
1	Parapet Wall			
2	Thermal Insulation			
3	Waterproofing			
4	Roof Tiles			
5	Installation of Drains			
6	Installation of Gutters			
Elevator System				
1	Installation of Rails			
2	Installation of Door Frames			
3	Installation of Cabin			
4	Cabin Finishes			
5	Installation of Wire Rope			
6	Installation of Drive Machine			
7	Installation of Controller			
HVAC Works				
1	Installation of Piping			
2	Installation of Ducting			
3	Installation of Insulation			
4	Installation of Dampers, Grills, and Diffusers			
5	Installation of Cladding			
6	Installation of Fans			
7	Installation of Fan Coil Units (FCU)			

(*Continued*)

TABLE 4.37 (Continued)
Checklist for Executed Works

Serial Number	Description of Works	Approved	Not Approved	Comments
8	Installation of Air Handling Units (AHU)			
9	Installation of Pumps			
10	Installation of Chillers			
11	installation of Cooling Towers			
12	Installation of Thermostat and Controls			
13	Installation of Starters			
14	Building Management System (BMS)			
Mechanical Works				
A. Water Supply				
1	Installing of Piping System			
2	Installation of Pumps			
3	Filter Units			
4	Toilet Accessories			
5	Installation of Insulation			
6	Water Heaters			
7	Hand Dryers			
8	Water Tank			
B. Drainage System				
1	Installation of Pipes (below grade)			
2	Installation of Pipes (above grade)			
3	Installation of Manholes			
4	Installation of Clean Out, Floor Drains			
5	Installation of Sump Pumps			
6	Installation of Gratings			
C. Irrigation System				
1	Installation of Piping System			
3	Installation of Pumps			
4	Installation of Controls			

(*Continued*)

TABLE 4.37 (Continued)
Checklist for Executed Works

Serial Number	Description of Works	Approved	Not Approved	Comments
D. Fire Fighting System				
1	Installation of Piping System			
2	Installation of Sprinklers			
3	Installation of Foam System			
4	Installation of Pumps			
5	Installation of Hose Reels/ Cabinet			
6	Installation of Fire Hydrant			
7	External Fire Hose Cabinet			
Electrical Works				
1	Conduit—Raceways			
2	Cable Tray—Trunking			
3	Floor Boxes			
4	Wiring			
5	Cabling			
6	Installation of Bus Duct			
7	Installation of Wiring Devices/Accessories			
8	Installation of Light Fittings			
9	Grounding			
10	Distribution Switch Boards			
Fire Alarm System				
4	Installation of Detectors, Bells, Pull Stations, Interface Modules			
5	Installation of Repeater Panel			
6	Installation of Mimic Panel			
7	Installation of Fire Alarm Panel			
Telephone/Communication System				
1	Installation of Racks			
2	Installation of Switches			
Public Address System				
5	Installation of Speakers			
6	Installation of Racks			

(*Continued*)

TABLE 4.37 (Continued)
Checklist for Executed Works

Serial Number	Description of Works	Approved	Not Approved	Comments
7	Installation of Equipment			
Audio Visual System				
5	Installation of Speakers			
6	Installation of Monitors/ Screens			
7	Installation of Racks			
8	Installation of Equipment			
CCTV/Security System				
1	Installation of Cameras			
2	Installation of Panels			
3	Installation Monitors/ Screens			
Access Control				
1	Installation of RFID Proximity Readers, Fingerprint Readers			
2	Installation of Magnetic Locks, Release Buttons, Door Contacts			
3	Installation of Panel			
4	Installation of Server			
Systems Integration				
1	Installation of Switches			
2	Installation of Servers			
External Works				
1	Site Works			
2	Asphalt Work			
3	Pavement Works			
4	Piping Works			
5	Electrical Works			
6	Manholes			
7	Road Marking			

Apart from project quality check/inspection, it is also required to assess the following activities on a regular basis to track project progress and its compliance with contract documents:

1. Schedule
2. Budget
3. Resources
4. Project Team Performance
5. Risk Assessment
6. Safety

Monitoring and control of schedule, budget, and resources (project progress) is performed on a regular basis by project monitoring and control team members. Also, team members from the supervision team as well as from the contractor side are assigned to carry out risk assessment and safety assessment. Performance evaluation is done by the line manager in coordination with the corporate office requirements.

4.6 CUSTOMER SATISFACTION

Quality is meeting customer requirements and customer satisfaction. Customer satisfaction for construction is delivering a project or facility as per the defined scope within the agreed-upon budget and specified schedule. It is delivering a project or facility that satisfies owner/end user needs and requirements. Customer satisfaction is an important factor in the development of the construction process and customer relationships.

Customer satisfaction can be used for the evaluation of quality and ultimately for the assessment of the success of an organization's quality improvement program and organizational focus and key in the evaluation of project team performance. Most organizations solicit comprehensive feedback regularly from their customers to share openly with the entire organization. The meaningful data and feedback concerning an organization's operating environment and the organization's performance in that arena, are essential for effective response and organizational success. However, to achieve a competitive advantage, meaningful data and feedback must be a continuous process. It needs to be as objective as possible to help identify areas needing improvement.

Construction is an increasingly competitive industry where organizational focus on exceeding expectations takes on added significance. It is important for improved performance and essential to establish a competitive advantage. To achieve a competitive advantage, a contractor should emphasize to produce high-quality construction facilities/projects. Customer satisfaction and continuous improvement are fundamental goals to achieve competitive advantage. Customer satisfaction enables construction companies to differentiate themselves from their competitors and create sustainable advantages.

In construction, a customer may be defined as the owner of the project or facility and the one that needs the construction facility (end user). Customer satisfaction in the construction industry is how well a contractor meets the customer's (owner's) expectations. Customer satisfaction in construction does not guarantee that future contracts will be contracted to the same contractor based on customer satisfaction criteria. Generally, a contractor's selection criteria are based on price but also on the contractor's technical and financial capability and previous experience of contractor's competence, however, the contractor's relationship and the level of satisfaction with the owner/client have significant importance to securing future jobs.

Customer satisfaction is therefore reflective of the customer's experience of and confidence in the contractor's abilities and cooperation. The essential objective in improving customer satisfaction is to achieve client loyalty, which can ensure that the contractor is considered and prequalified for future contracts or partnering guarantees.

Customer satisfaction of construction is a function of several factors such as:

- Customers' past or direct experience with the contractor.
- References about the contractor having successfully completed projects of similar nature.
- Customer's personal needs which contractor can satisfy.
- Contract deliverable system.
- Bidding procedures and requirements.

Construction projects are of temporary endeavor, unique having finite duration. Customer satisfaction enables construction companies to differentiate themselves from their competitors and create sustainable advantages. Construction projects involve numerous stakeholders that are closely related and interact during a given project. The level of stakeholders' satisfaction directly influences the current project and subsequent projects and the level of satisfaction experienced by the stakeholders. For contractors, completing a project in accordance with the defined scope within budget and on time satisfies the owner needs and generates profit. The relationship between customer and contractor is periodic and depends on the duration of the project. In construction, mutual cooperation, and maintaining a harmonious working relationship between customer (owner) and contractor is strongly emphasized. Further, in construction, the project organizations and collaborative relationships are of a "one-off" nature, the customer satisfaction survey model is different from those in other industries where the relationship between the client (customer) and the supplier is normally a long-term relationship. In construction, each customer has unique needs and expectations, therefore, make a concerted effort to identify customer satisfaction through direct discussions with the owner and participation in meetings with the owner and their team prior to the start of the project. Therefore, while creating customer satisfaction surveys for construction the following primary objectives should be considered:

- Establishing client expectations.
- Identify the most effective methods used to solicit customer feedback.
- Procedure to collect the feedback.
- Investigate how customer feedback is utilized.

An example model of a customer satisfaction survey in construction is illustrated in TABLE 4.38.

TABLE 4.38
Customer Satisfaction Survey

Sr. No.	Stages	Elements	Description
1	Preconstruction	Establish client expectation.	Contractor to have one-on-one discussion with owner/client, project manager, consultant (construction supervisor) prior to start of construction to help: 1. Establish customer expectations. 2. Determine performance criteria. 3. Define team expectations. 4. Determine required team members.
2	Construction	To know how the work is progressing.	Conduct meetings at regular intervals to find out whether: 1. Client's expectations are met. 2. Work progress satisfactory: a. Schedule b. Material c. Workforce 3. Specified quality is achieved. 4. Cost variance. 5. Any discrepancies with documents. 6. Staff performance and cooperation. 7. Subcontractor's performance.
3	Post Construction	Prepare survey questionnaire and get feedback from owner/client, project manager, consultant.	The survey questionnaire to include items related to the following factors: 1. Quality assurance at work: 1.1 Management and implementation of agreed-upon quality assurance procedures. 1.2 Quality of material. 2. Workmanship: 2.1 Quality of work. 3. Skill of personnel. 4. Cooperation of site staff:

(Continued)

TABLE 4.38 (Continued)
Customer Satisfaction Survey

Sr. No.	Stages	Elements	Description
			4.1 Agreement about changes/variations.
			5. Behavior of site staff.
			6. Work supervision.
			7. Attendance to defects.
			8. Safety at work:
			8.1 Safety measures.
			8.2 Handling of accidents.
			8.3 Attendance to injuries/accidents.
			9. Environmental consideration:
			9.1 Cleanliness of site.
			9.2 Handling of hazardous material.
			9.3 Disposal of waste.
			10. Subcontractor's performance:
			10.1 Competency of subcontractor(s).
			11. Adherence to schedule.
			12. Overrun.
			13. Information flow.
			14. Handover procedure.
			15. Attending snags.
			16. Overall satisfaction.
			17. Consideration for future projects.

Implementation of customer satisfaction surveys helps manage and improve customer relationships. In addition, measuring customer satisfaction has several benefits for organizations. These are:

- Improvement in communication between client and contractor/supplier.
- Improvement in mutual understanding.
- Identify the areas for improvement.
- Better understanding of the problems.
- Evaluation of progress towards the objectives.
- Monitoring accomplishments.
- Monitoring changes.

Customer satisfaction in the construction industry is different from that of manufacturing, process, and services industries, where several evaluation models or indices exist for accessing customer satisfaction surveys and collecting data to correct or prevent defects/mistakes. The function of the construction industry is to provide customers with facilities that meet their needs and expectations. The data collected through customer satisfaction questionnaires in the construction industry can only be used to improve the next product/project. TABLE 4.39 lists customer satisfaction questionnaires that can be used to get related information about completed building construction projects.

TABLE 4.39
Customer Satisfaction Questionnaires

Sir/Madam

We are conducting a survey to improve our performance. We would appreciate your response to the following questions. Your valuable information will help us to satisfy your requirements in a better way. (The information provided by you shall be kept private.)

Customer Satisfaction Questionnaire

Name of Organization:

Name of Completed Project:

Q1 Workmanship of the following installed items

Sr. No.	Work Description	Very Good	Good	Average	Poor	No Comment
1	Structural Work					
2	Precast Work					
3	Internal Finishes					
4	External Finishes					
5	Plumbing					
6	Drainage					

(Continued)

TABLE 4.39 (Continued)
Customer Satisfaction Questionnaires

7	Fire Fighting					
8	HVAC					
9	Elevator					
10	Escalator					
11	Electrical Work					
12	Landscape Work					
13	Pavements					
14	Streets/Roads					
15	Water Proofing					
Q2	**How are the following Systems functioning**					
Sr. No.	**Systems**	**Very Good**	**Good**	**Average**	**Poor**	**No Comment**
1	HVAC					
2	Water System					
3	Lighting					
4	Power					
5	Fire Alarm System					
6	Telephone System					
7	Public Address System					
8	Security System					
9	Kitchen Equipment					
10	Window Cleaning System					
11	Irrigation System					
12	Elevators					
13	Escalators					

(Continued)

TABLE 4.39 (Continued)
Customer Satisfaction Questionnaires

Q3	**How was the behavior of our site staff**			
Sr. No.	**Position**	**Co-operative**	**Polite**	**Unpolite**
1	Managers			
2	Engineers			
3	Foreman			
4	Labors			
5	Office staff			
Q4	**How was site safety**			
	Good ☐	Average ☐	Poor ☐	
Q5	**Whether project was completed as per schedule**			
	Yes ☐	No ☐		
Q5A	**If your reply to Q5 is No, then please let us know:**			
	Have you incurred any revenue loss/any other damages due to non-completion in time?			
	Yes ☐	No ☐		
Q6	**How was the response to your problems at site**			
	Immediate ☐ Within one week	☐ Within two weeks	☐ No response	

(*Continued*)

TABLE 4.39 (Continued)
Customer Satisfaction Questionnaires

Q7	**Whether our variation claims were as per contract**					
	Yes ☐	No	☐			
Q8	**Are you satisfied with variation settlement method**					
	Satisfied ☐	Not satisfied	☐	Needs improvement	☐	
Q9	**Are you satisfied with our performance**					
	Satisfied ☐	Not satisfied	☐	Needs improvement	☐	
Q10	**How was the response to your problems from our Head Office**					
	Good ☐	Poor	☐	Needs improvement	☐	
Q11	**How do you compare our performance compared with other competitors**					
	Better than others ☐	Same as others	☐	Less than others	☐	
Q12	**Whether you will consider us inviting for your new tenders**					
	Yes ☐	Can't say	☐			
Q13	**Any Suggestion(s)**					

5 BIM in Design and Construction

Cliff Moser

5.1 INTRODUCTION

5.1.1 What Is Building Information Modeling (BIM)

Building Information Modeling (BIM) is not just the design and construction modeling of buildings but the entire process of designing and managing the digital three-dimensional (3D) representations of a proposed or existing built space. BIM is an update of computer-aided drafting (CAD) and a new relationship with design and construction. BIM 3D components contain information and are located within the model's 3D space. If modeled correctly, digital views or slices of the model will show not just plan, section, and elevation views but also fine details of the component and BIM building files. BIM components are also parametric, meaning they are shaped by 3D and data algorithmic processes that tie them to location and relationship with other BIM components. Changing one BIM component can automatically change the attributes of another BIM component. For example, BIM components such as partitions, electrical conduits, ductwork, and fixtures can automatically adjust their relationships with other components as their attributes are modified. These adjustments are bidirectional, so components that are modeled correctly will change other components in the model as required.

Another way BIM differs from CAD is that BIM files contain digital information about each of its separate components and the entire building, such as materials, finishes, and equipment requirements like electrical voltage requirements. This information makes up the "I" in BIM, creating a searchable database of the model. This data is additive and scalable, enabling design and construction team members to participate in the virtual co-creation and co-development of a project's design and operational requirements. BIM facilitates 3D digital development, including coordination, estimating, scheduling, and material selection. BIM also creates a rich data source for rules-based project studies, such as occupancy simulations, fire and life-safety code reviews, and energy use calculations. BIM has been used since the 1990s, and its diffusion into the collaborative design and construction processes has changed the industry. Therefore, BIM is used for all construction projects—large and small. Most, if not all, designers and constructors use BIM in their project deliverables.

As its use is so widespread, this chapter will identify ways to utilize BIM to support quality in the design and construction industry.

DOI: 10.1201/9781003636687-5

5.2 BIM AS COLLABORATIVE QUALITY TOOL

5.2.1 Sharing Information

The design and construction industry drives economies. Its international value in 2024 is estimated at $10.3 trillion US dollars. By 2030, that number may be $14.4 trillion.[1] Regional and national governments regularly use construction to increase employment, which increases GDP while building long-term public and private infrastructure. However, most of its participants are comprised of small, low-profit firms, which temporarily collaborate on projects that often suffer serious setbacks, such as schedule and budget overruns, resulting in litigation, claims, and organizational defaults.

Furthermore, the traditional ways that projects are contracted exponentially facilitate project and organizational failure. Most projects are still typically contracted as a design-bid-build (DBB), arranged through tri-partite contract agreements, in which the owner holds separate contracts with the design and building team. These arrangements typically identify the design team as the project lead in providing the project's design, code, and owner's requirements. Once the design is sufficiently complete, the design team delivers its construction documents to the owner, who puts the project out to bid to contracting teams. Once a contracting team is selected, this team begins to build the project, usually with multiple trade and subcontractors and multiple vendors and suppliers. This tripartite contractual arrangement makes the builder and designer dependent on each other's work product as a third party, it also creates adversarial relationships. If the builder believes that the designer delivered incomplete or deficient documents to the owner, then the builder must petition the owner to demand new information from the designer. The designer typically responds that its documents are complete and it is the builder who is delivering deficient services.

In this traditional design and construction role, the owner, as the holder of both contracts, must find ways to ensure that I) the design team delivers a work product that matches their contract and legal liability and standard of care requirements and II) that the builder knows its requirements and has the skill and expertise to deliver the project successfully. BIM allows the owner to validate these requirements by creating the BIM file execution and sharing requirements before completing the design team's files and delivering them to the builder.

These requirements are formalized in the BIM Execution Plan (BEP).[2] The BEP delineates the file-sharing requirements and responsibilities for each team representing the roles of owner, design team, and builder. If the owner develops this plan before embarking on its BIM journey, it can help define deliverable requirements and progress milestones for its design and construction teams. A BEP can be used for other types of design and construction contracts. Such as design-build (DB)[3]—in which the builder or the design team holds the contract with the other party. The design-builder team has a single contract with the owner in this agreement. The BEP can also be used for different types of agreements like the Integrated Form of Agreement (IFOA) and Integrated Project Delivery (IPD).[4] These agreements designate a single project-team entity through a single-signed team contract, in which all project team members, owner, builder, and design team sign together.

5.2.2 Managed BIM File-Sharing

The BEP formalizes BIM file sharing to an agreed set of milestones and deliverables. It also formalizes a series of decided development and transfer protocols. Described as BIM Level of Development (LOD),[5] the LOD outlines the component and model information, data, and attributes necessary for the model to be accepted to a quality standard before being transferred to another team member. By defining recognized industry standards, LOD ensures that the design models have the right amount of information in them, protecting them from disputes regarding incomplete information before being transferred to the builder. LOD, combined with a BEP, provides the teams with a standard of care design intent, construction means, and liability and intellectual property control methods by defining how the models will be used as they are transferred between separately contracted team members.

5.2.3 Team Collaboration

Managed BIM file sharing enables team quality-focused coordination and completion of shared team BIM files. Starting with the BEP accepted transferred LOD design files, the trade contractors like the Mechanical, Electrical, and Plumbing (MEP) trades can begin their coordination and detailing activities. This process involves adding additional LOD information to the BIM files and coordinating with the other design and construction team members. Since correctly modeled BIM components include location attributes, these teams aggregate their files by coordinating the systems against each other, adjusting their files as allowed by their contracts and construction means and methods. The design teams are involved in this process by reviewing the proposed revisions against their design and code requirements. These coordinated BIM files also identify the construction sequence and materials needed to complete construction. Typically, in the past, these deliverables were two-dimensional non-digital plans overlaid over the design teams' design drawings. This coordination was not a collaboration. Collaboration involves not just the sharing of BIM files but also the sharing of expertise. Design intent can benefit from contractor means and methods early into the beginning of the project.

5.2.4 Big Rooms

A big room is a collaboration and quality system that was first widely adopted by Toyota Motor Company. It was famously used to develop the Prius automobile.[6] It soon became a signature Toyota process to ensure that all stakeholders were involved in developing and selling a new and innovative automobile. Engaging an entire team, including all stakeholders, ensures that every aspect of the production process is carefully reviewed and accepted.

In design and construction projects, the big room process ensures that all stakeholders participate within a single collaborative space. Invite the building team of general and trade contractors and their major vendors and suppliers, as well as all members of the design team, including the MEP engineers and interior designers. The owner's teams participate as well, including inspectors and the finance team.

A big room creates a shared "BA" space[7] for directed project learning, with a focus on the project's mission, with everyone understanding, embracing, and promoting the goals of the entire project. By including everyone in their Prius big room, Toyota ensured that the design, manufacturing, and sales goals could be met.

A construction big room should have the same goals—not only to design and build the project but to achieve something that can only be accomplished by completing this project. When I would work with a team on a hospital design and construction project, our goal was singular—"When finished, this Project will change the face of Healthcare." By projecting and embracing this goal, the team would understand that their role and participation were necessary to change the world. Required participation in a collocated, collaborative big room reinforces that mission. BIM enables that collaboration. A big room is typically set up during the pre-construction phase when the building team is just coming on board to support and augment the owner and design teams, and will be managed through the BEP. The BIM construction coordination and detailing activities drive the big room conversations regarding design, schedule, and cost. As team members begin to see the project virtually taking shape inside the virtual world of BIM, they begin to imagine how their role and effective engagement guide the project's development.

The big room should continue into construction and only be disbanded after project activation, as the owner's facilities team is taking over operations of the building. The BIM is also necessary for this role as it is developed into an operations and facilities model (LOD 500+), with the construction as-builts and the design team record models available for the owner's teams.

5.3 QUALITY SYSTEMS, QUALITY ASSURANCE, AND QUALITY CONTROL

5.3.1 Quality Systems

Before the project starts, the owner should develop a quality system that defines the BIM deliverables based on LOD and the BEP. Once created, the design team should share and update it as they begin working on the project. Once the builder begins their work, the quality system should guide their deliverables. This arrangement ensures that all the team members understand and agree with the quality system and standards jointly developed by the project team. This system could include a quality assurance and a quality control program that follows accreditation and certification requirements like ISO 19560.[8] ISO has a library of quality standards for design and construction.[9] Suppose the project team doesn't support ISO requirements. In that case, the project can still benefit by understanding the requirements for a robust and resilient quality system and just being aware of the rigor that ISO brings to quality can benefit both the team and the project. The BEP can be modified to include project-specific quality requirements.

5.3.2 BIM 2D to 10D

The LOD rating also guides how BIM files can be used beyond 3D coordination. Known in the AECO industry as the "Ds," they identify how BIM files are used by third-party software.[10] The Ds are additive, with 2D BIM stripped of much of its information and flattened into 2D CAD or PDF files. 3D is BIM in its native format, used by the teams coordinating BIM files for design and construction, and shared through common data environments. 4D is BIM used for scheduling and adding time to the 3D files. 5D adds cost to the time component of 4D BIM. 6D adds life-cycle costing to the 3, 4, and 5Ds. 7D adds sustainability and green building information to the 4, 5, and 6D components. 8D adds construction safety to the 3, 4, 5, 6, and 7Ds. 9D supports aspects of Lean construction and design, and 10D endorses the industrialization of the industry with manufacturing. Software supporting the Ds continues to be created year after year. Knowing how a model will be used can help teams develop complete and accurate BIM files.

5.3.3 Quality Assurance

Quality Assurance (QA) on a design and construction project is relevant for all project team members. Applying QA to the design team's professional services would include activities established by the owner and code and standard of care requirements. Design team QA comprises an internal review of the deliverables against these requirements to ensure that the project advances toward its contracted deliverables and how models are delivered. This metric would be a service quality measure. Typically, QA activities and responsibilities are delineated within the quality system, while quality control (QC) is a subset of the QA activities.

5.3.4 Quality Control

In professional services, QC should guide the delivered project results and the service in which that product is delivered. The design team develops BIM files defined to the owner's requirements and provides them against their professional standard of care. That review becomes another QC check when the files are shared with the owner's and builder's team.

For design and construction projects in the past, these QA/AC processes were enforced through inspections, both internal and external. The design team performed internal deliverable inspections before delivering their documents to the owner. Like LOD with BIM, 2D contract drawings would be audited against milestone completion requirements, including program and Schematic Design (SD) documents, which the owner reviewed and, when approved, transferred to the design team's internal production group, which would continue to develop Design Development (DD) and Construction Documents (CD).

In this system, completed design deliverables were internally measured by quality control, accepted within defined tolerances, and then released for use to the owner, who, in a design-bid-build contract, would release the documents for contractor bidding. In this scenario, the design team's quality assurance of its documents would be audited by the builder.

Once received by the builder, deficiencies discovered in the documents were raised through Requests for Information (RFIs),[11] often resulting in change orders that amended the builder's contract with the owner. Frequently, the design team's deficient QC resulted in the need for the design team to revise their documents and for the builder to demolish and reinstall work already completed in the field. Because an inspection-based system supported the design team's quality control, it relied on an apprenticeship methodology that paired experienced project architects with junior staff trained through an iterative development of drawings and details and continuous experienced architect review and oversight. This is the apprenticeship system I was trained under with the result being an apprentice gradually learning not only the requirements of the project but the firm, the industry, and the profession. These apprenticeship processes are still built into the profession's licensure requirements, as emerging graduate architects still need several years of experience under the supervision of a licensed architect before they can sit for the Architectural Registration Exam (ARE).[12] This period is formalized as the Architectural Experience Program (AXP).[13] This program was formally called the Intern Development Program (IDP) and was based on professional knowledge being taught through personally experienced oversight rework, similar to craftsman-apprenticeship activities in the past.

However, with the advent of electronic deliverables such as CAD and now with BIM, there is a considerable gap in the oversight and rework process (which is one of the reasons the IDP was revised into the AXP). Master draftsmen no longer exist in design firms, and hard-copy review oversight of design team construction documents before delivery is impossible, as no one internal design team member or group can provide individual design team oversight.

5.4 PROCESS-BASED SYSTEMS

The AXP recognizes that the design profession needs to develop its QA/QC systems on process-based quality, not inspection-based quality, to improve the processes for developing and providing services for the project's required deliverables. Process-Based systems are most evident in the Lean design and construction process. Based on the Toyota Production System (TPS),[14] Lean is based on production quality—correct and complete deliverables with sample conformance audits post-delivery. In Lean, inspection of deliverables and services is used to confirm that the production system is working. Lean ensures that requirements are understood and met for design and construction projects and that the entire team is actively engaged in defining and delivering the project as responsible agents. This is achieved through team training, education, and management support.

5.5 MACHINE LEARNING QUALITY CONTROL WITH BIM

BIM enables the requirements and delivery of project requirements to be digitally measured. Large and Small Language Models (L/SLM),[15] which use large, aggregated amounts of data or small specific industry or project-type data (still large but confined to a particular industry), can be used to train project-specific machine learning through Generative Pre-Trained Transformers (GPT).[16] GPTs have been successful for other industries that rely on developing predictive and reporting models through queries from mining from large amounts of information; as these systems become more mainstreamed for construction projects, it is only a matter of time before a GPT application can audit the design and construction BIM files, at various stages of LOD, to confirm that the project deliverables meet the project requirements. This GPT could develop and present a variance report listing discrepancies between the design and deliverables and discrepancies within domains such as architecture and engineering models. A GPT could also offer solutions to the proposed changes, pricing, and scheduling discrepancies. GPTs or other Natural Language Processing (NLP)[17] tools could also create templated BIM files from the owner's requirements as starting points for the design team. For example, a conceptual or schematic BIM (LOD 100 to 200) could be developed by the GPT or another machine learning application before the design team even becomes contracted to the project. This would entail creating a starting BIM building a model from the owner's requirements, basis of design, and fire and life safety codes. We see these efforts beginning as NLPs modify the BIM for room dimensioning.[18]

The future of using these new tools argues that a robust process-based quality program should be instituted to support and guide the continued blended AI, owner, design, and construction team development of BIM files.

5.6 OTHER AI BIM TOOLS

Utilizing LLM, AI has the ability to design and coordinate a building as well. Using reference design requirements, AI can develop multiple schematic ideas from which the design and owner's teams can select. AI can measure the LOD of BIM files. Already, design teams are also being used to help develop the Ds of BIM. The 4D attribution functions of BIM are being used by some firms, such as Alice Technologies,[19] to create construction schedules. ProEst[20] is an app that generates estimates from the BIM files. Dusty Robots[21] uses BIM files for autonomous field layouts.

5.6.1 Point Cloud Files

As we see from AI, creating a BIM file from scratch is only one of multiple ways to start a design and construction project. If the project is a remodel, renovation, or even a new building to be inserted between existing buildings, the project team may want to start modeling based on the results of a point cloud file. Point cloud files are 3D representations of an existing physical space, like site surveying files. These files are created by capturing billions of digital data points reflected by a laser scanner, processed through a Light Detection and Ranging (LiDAR)[22] system of high-resolution points. Because point cloud data is 3D, it contains information for the X, Y, and Z axes and is located geo-spatially.[23] Since the data is graphic and dense, a "cloud" of points is represented in a point cloud model. As with most digital files, this information can be transferred and processed into a BIM file. Once received, the modeling teams can add data and attributes to build a model to be developed and shared with the project's teams.

As native files, point cloud files have limited "smart" information and depend on the BIM to modify the information into intelligent information for sharing or building new models. These BIM files can then be used to begin design studies and develop construction documents as further LOD information is added. BIM can be added to Geographic Information Systems (GIS) to map the building to the site.

5.7 DIGITAL TWINS

Another innovation that is changing the AECO industry through BIM is the digital twin.[24] A digital twin is just what it sounds like—the digital version of the physical building. A digital twin can start with an owner's existing BIM or serve as a template for a new twin to be initially developed by the design team. Once complete, the digital twin can become an integrated facility and operations BIM (LOD 600) linked and interoperable with the existing physical structure. This interoperability is managed through facility operations systems such as Maximo,[25] which provides equipment service interoperability, and Archibus,[26] which provides workplace management tools such as occupancy and asset management. With a digital twin, as the owner's equipment or occupancy requirements are added or modified (in both the physical or digital space), the digital and physical twins conform and update each other. The design team creates digital twin as if they are beginning a new file, with conceptual or schematic design modeling (LOD 100 to 200). As they progress through the development, they add additional data to ultimately connect it to the owner's facility and operations system. Interoperability is necessary between the physical built environment and the digital BIM. This is achieved through the Internet of Things (IoT) of sensor-connected hardware and systems. These systems are overlaid during the design and construction BIM development and commissioned during owner activation and turnover.

5.8 PROJECT COMPLEXITY

Design and construction projects are Complex Adaptive Systems (CAS).[27] Multiple interacting activities and agents define a CAS. These conditions facilitate the ability of processes to adapt their behavior and roles based on the project environment and their interaction with other activities and agents. CASs present unique conditions such as ad hoc hierarchies, group emergence, and self-organization, enabling singular responses to shifting project conditions. CASs exist throughout interconnected systems such as transportation systems and financial markets.

Furthermore, the tripartite relationships between the owner, design, and builder teams create temporary multilayer networks[28] in which actions in one layer can exponentially propagate and cascade into other layers. For example, a discrete action in one layer of a connected multilayer network, like the owner withholding payment to the builder or the design team changing a material specification, can cascade into unforeseen changes in other layers, affecting schedule and cost beyond what original activity anticipated. Infection networks are examples of multilayer networks. Using BIM to communicate and monitor progress and activities within the other layers could help teams understand and track unanticipated actions that may cause disruptions in other project layers. As one can see, multilayer networks CASs are challenging to manage, as they respond differently to control or actions. Like the withholding payment example above, attempts to control individual or group behavior result in unexpected outcomes, which quickly cascade and propagate in unknown ways. Project complexity also involves agent emergence and self-organization. As the multiple team layers work in their particular domains producing BIM files to be transferred into construction, traditional hierarchical reductionist project management[29] becomes ineffective. The project's multilayer agents will find ways to work around these project controls and self-organize into project-based holons.[30] As holons, they create their own project controls, which they use to guide the projects adaptively. Holons are temporarily formed teams consisting of agents from each project multilayer, who emerge and self-organize to manage both their domain layer (design, construction, or owner layer) and the project needs. Holons are delineated as having two faces, the domain face—the holon's discipline expertise, and the project face—role the holon is playing on the project. Holons develop and adapt as required between these sometimes-connected layers.[31] They are developing systems that nurture and support the emergence of holons.

IPD and IOFA project agreements provide a fertile environment for creating holons, but since all construction projects are complex systems, holons will develop in any contract environment. Making traditional DBB/DB project agreements flexible with selective Time and Material (T&M)[32] processes enables more responsive controls that encourage and support holonic development. I've used selective T&M to frame the development of holons for BIM and construction tasks within a larger DBB contract. T&M facilitated the formal emergence of holons to plan, BIM, and construct defined portions of the project. This development allowed the use of Lean construction tools such as the Last Planner System[33] to be implemented, where it had been resisted in the larger project. T&M created a payment program that included the design team, which charged their time to the defined T&M project tasks. T&M helped create shared goals and incentives that guide complex adaptive systems.

5.9 CONCLUSION

More than a single tool or process, BIM continues to transform the systems of design and construction. It can help define and integrate industry processes, such as contracts, quality systems, code, and standard of care requirements, and digitally connect buildings and systems. BIM LOD and the "Ds" enable the ability to measure the development of models and integrate systems of schedule, cost-estimating, sustainability, and even safety into the design. Files can be easily shared between different team members and AECO domains such as design, construction, and owner's teams. Autonomous model development with AI tools can help ensure code and owner requirements. Existing condition models can be created from point clouds, and BIM-enabled digital twins can ensure that the physically built environment continuously matches the virtual built environment.

NOTES

1 https://www.statista.com/statistics/1290105/global-construction-market-size-with-forecasts/
2 https://www.nibs.org/nbims/bep/v4
3 https://dbia.org/what-is-design-build/
4 https://www.consensusdocs.org/contract-series/300-collaborative-agreements/
5 https://bimforum.org/resource/lod-level-of-development-lod-specification/

6 https://ideas.repec.org/p/hit/hjbswp/106.html
7 https://www.researchgate.net/figure/Different-forms-of-learning-space-ba-Nonaka-Konno-1998_fig2_222825255
8 https://www.bsigroup.com/en-US/bim-iso-19650/
9 https://www.iso.org/publication/PUB100317.html
10 https:// https://constructionblog.autodesk.com/bim-dimensions//
11 https://essential.construction/academy/tutorials/division-01-writing-an-effective-rfi/
12 https://www.ncarb.org/pass-the-are
13 https://www.ncarb.org/gain-axp-experience
14 https://global.toyota/en/company/vision-and-philosophy/production-system/
15 https://arxiv.org/abs/2305.08848
16 https://en.wikipedia.org/wiki/Generative_pre-trained_transformer
17 https://www.ibm.com/topics/natural-language-processing
18 https://www.archsmarter.com/blog/revit-api-chatGPT
19 https://www.alicetechnologies.com/home
20 https://proest.com/
21 https://www.dustyrobotics.com/
22 https://www.usgs.gov/faqs/what-lidar-data-and-where-can-i-download-it
23 https://www.ibm.com/topics/geospatial-data
24 https://ascelibrary.org/doi/10.1061/9780784483961.070
25 https://www.ibm.com/products/maximo
26 https://archibus.com/
27 https://leanipd.com/blog/is-the-construction-industry-a-complex-adaptive-system/
28 https://www.dustyrobotics.com/
29 https://journals.sagepub.com/doi/full/10.1177/14761270211038635
30 https://www.sciencedirect.com/science/article/abs/pii/S0926580518309968
31 https://www.manacontemporary.com/editorial/a-brief-primer-on-holons-and-holarchy/
32 https://www.projectmanager.com/blog/time-and-materials-contract
33 https://leanconstruction.org/lean-topics/last-planner-system/

6 Construction Contract Documents

Abdul Razzak Rumane

6.1 INTRODUCTION

Construction is translating the owner's goals and objectives, by the contractor, to build the facility as stipulated in the contract documents, plans, and specifications within the schedule and budget.

Construction projects have the involvement of three main participants. These are:

1. Owner
2. Designer/Consultant
3. Contractor

In certain cases, owners engage a professional firm, called a construction manager, who is trained in the management of construction processes, to assist in developing bid documents, overseeing, and coordinating projects for the owner or use a project manager type of contract when the owner decides to turn over entire project management to a professional project manager. Participation/involvement of all three parties at different levels of construction phases varies depending on the type of project delivery system.

Construction contract documents are written documents that constitute a set of contract documents developed to meet the owner's needs, required level of quality, schedule, and budget. Based on the type of project delivery system and type of contracting arrangements, necessary documents are prepared by establishing a framework for the execution of a project to satisfy owner requirements. It is essential that contract documents are clearly written in simple language that is unambiguous and convenient to understand by all the concerned parties. These documents mainly consist of the following:

1. Bidding
    - Tendering Procedures
    - Bid Bond
    - Performance Bond
    - Agreement Forms

2. Contracting
    - General Conditions
    - Particular Conditions

DOI: 10.1201/9781003636687-6

3. Construction
    - General Specifications
    - Particular Specifications
    - Drawings
    - Bill of Quantities
    - Price Analysis Schedule
    - Forms
    - Reports,
    - Soil Test Reports
    - Survey Reports
    - Addenda (if any)
    - Special Technical Requirements

There are organizations/agencies producing different types of construction contract formats that are globally used as a guide to prepare construction contract documents. The following are the best-known organizations whose documents are most commonly used to prepare construction contract documents for a specific project:

1. EJCDC®: The Engineers Joint Contract Documents Committee, USA
2. FIDIC: Federation Internationale Des Ingenieurs (International Federation of Consulting Engineers)
3. MasterFormat®: Construction Specifications Institute (CSI) and Construction Specifications Canada (CSC)
4. NEC: New Engineering Contract (NEC) or NEC Engineering and Construction Contract—UK Institution of Civil Engineers

6.2 EJCDC

The Engineers Joint Contract Documents Committee (EJCDC®) is a joint venture of the American Council of Engineering Companies (ACEC), the National Society of Professional Engineers/Professional Engineers in Private Practice (NSPE/PEPP), and the American Society of Civil Engineers-Construction Institute (ASCE-CI).

EJCDC® has existed since 1975 to develop and update fair and objective standard documents that represent the latest and best thinking in contractual relations between all parties involved in engineering design and construction projects. EJCDC® represents a major portion of the professional groups engaged in the practice of providing engineering and construction services for the construction project.

Each EJCDC® contract document is prepared by experienced engineering design and construction professionals, owners, contractors, professional liability and risk management experts, and legal counsel.

The following is the list of EJDCD® contract document categories:

1. Joint Venture Documents
2. Environmental Remediation Documents
3. Construction-Related Documents
4. Funding Agency Documents

5. Procurement Documents
6. Design-Bid-Build Documents
7. Guides, Commentaries, and References
8. Design-Build Documents
9. Owner Engineer Documents
10. Peer Review Documents
11. Engineer Sub-Consultant Documents

FIGURE 6.1 illustrates the table of contents of Standard General Conditions of the Construction Contract published by EJCDC®.

More information about the EJCDC® contract documents is available at www.ejcdc.org.

6.3 FIDIC

Fédération Internationale des Ingénieurs Conseils (from French International Federation of Consulting Engineers) "FIDIC" was founded in 1913 by Belgium, France, and Switzerland. FIDIC now has about 78 member associations from all over the world.

The first edition of the *Conditions of Contract (International) Works of Civil Engineering Construction* was published in August 1957 having been prepared on behalf of FIDIC and the Fédération Internationale des Bâtiment et des Travaux Publics (FIBTP). The form of the early FIDIC contracts followed closely the fourth edition of the (Institute of Civil Engineers)—*ICE Conditions of Contract.*

FIDIC publishes international standard forms of contracts for works, and for clients, consultants, sub-consultants, joint ventures, and representatives, together with related materials such as standard prequalification forms.

FIDIC is well known for its international forms of contracts which are used worldwide by the construction industry and also endorsed by multilateral development banks (MDBs).

6.3.1 The FIDIC Suite of Contracts

In 1999, FIDIC published a revised suite of contracts. These books consist of two parts. The first part consists of general principles that should be included in the contract, while the second part explains how the employer can prepare the special conditions in accordance with the first part, based on the nature of the work in each construction contract.

The FIDIC Suite of Contracts includes the following books:

1. The Green Book—*Short Form of Contract*
2. The Red Book—*Conditions of Contract for Construction for Building and Engineering Works Designed by the Employer* (construction contract)
3. The Pink Book—*Harmonized Red Book* (MDB Harmonized Edition), *Conditions of Contract for Construction for Building and Engineering Works designed by the Owner*

This document has important legal consequences; consultation with an attorney is encouraged with respect to its use or modification. This document should be adapted to the particular circumstances of the contemplated Project and the controlling Laws and Regulations.

STANDARD GENERAL CONDITIONS
OF THE CONSTRUCTION CONTRACT

Prepared by

ENGINEERS JOINT CONTRACT DOCUMENTS COMMITTEE

and

Issued and Published Jointly by

SAMPLE

AMERICAN COUNCIL OF ENGINEERING COMPANIES

ASSOCIATED GENERAL CONTRACTORS OF AMERICA

AMERICAN SOCIETY OF CIVIL ENGINEERS

PROFESSIONAL ENGINEERS IN PRIVATE PRACTICE
A Practice Division of the
NATIONAL SOCIETY OF PROFESSIONAL ENGINEERS

These General Conditions have been prepared for use with the Suggested Forms of Agreement Between Owner and Contractor (EJCDC C-520 or C-525, 2007 Editions). Their provisions are interrelated and a change in one may necessitate a change in the other. Comments concerning their usage are contained in the Narrative Guide to the EJCDC Construction Documents (EJCDC C-001, 2007 Edition). For guidance in the preparation of Supplementary Conditions, see Guide to the Preparation of Supplementary Conditions (EJCDC C-800, 2007 Edition).

EJCDC C-700 Standard General Conditions of the Construction Contract.

7/17/07 {MW000852,1}

FIGURE 6.1 Table of contents (EJCDC® condition of construction contract).

STANDARD GENERAL CONDITIONS OF THE
CONSTRUCTION CONTRACT

SAMPLE

TABLE OF CONTENTS

EJCDC C-700 Standard General Conditions of the Construction Contract.

7/17/07 {MW000852.1}

FIGURE 6.1 (Continued)

SAMPLE

EJCDC C-700 Standard General Conditions of the Construction Contract.

Page ii

7/17/07 {MW000852;1}

FIGURE 6.1 (Continued)

SAMPLE

FIGURE 6.1 (Continued)

SAMPLE

FIGURE 6.1 (Continued)

4. The Yellow Book—*Conditions of Contract for Plant and Design-Build for Electrical and Mechanical Plant and for Building and Engineering Works, Designed by the Contractor* (plant and design-build contract)
5. The Orange Book—*Conditions of Contract for Design-Build and Turnkey*
6. The Silver Book—*Conditions of Contract for EPC/Turnkey Projects* (EPC turnkey contract)
7. The Gold Book—*Conditions of Contract for Design, Build and Operate Projects* (design-build-operate contract)
8. The Blue Book—*Form of Contract for Dredging and Reclamation Works*
9. The White Book—*FIDIC* Client/Consultant Model Services Agreement
10. The Conditions of Subcontract for Construction—*Conditions of Subcontract for Construction for Building and Engineering Works Designed by the Employer*

6.3.1.1 The Green Book—'Short Form of Contract'

This is FIDIC's recommended form of contract for use on engineering and building work of relatively small capital value or where the construction time is short. FIDIC has provided guidance that this would probably apply to contracts with a value of less than US $500,000 or a construction time of less than six months.

However, the Green Book may also be suitable for simple or repetitive work. Although typically in these types of arrangements the contractor is to construct the work according to the employer's design, it does not matter whether the design is provided by the employer or by the contractor. There is no engineer, and payments are made in monthly intervals.

The contents of the book are as follows:

- Agreement
- General Conditions
- Rules for Adjudication
- Notes for Guidance

The Green Book is a flexible document containing all the essential administrative and commercial arrangements. It is possible to easily amend and supplement the provisions of the Green Book with differing options incorporated via the appendix. The Green Book is likely to be the most suited, fairly simple for repetitive work or work of short duration without the need for specialist sub-contracts.

6.3.1.2 The Red Book—'Conditions of Contract for Construction for Building and Engineering Works Designed By the Employer' (The Construction Contract)

In 1956 the first edition of the original form of Red Book was published. This was updated and amended over four editions. Then in 1999, FIDIC released a totally new set of standard forms of contract, including a brand-new version of the Red Book, which superseded the original version of the Red Book. The Red Book is the FIDIC-recommended form of contract for building or engineering work where the employer has been responsible for nearly all the design. In fact, FIDIC claims that the Red

Book is the most widely used international construction contract where most of the work has been designed by the employer.

A key feature of the Red Book is that payment is made according to bills of quantities (i.e., a document in which materials, plant, and labor (and their costs are itemized), although payment can also be made on the basis of agreed lump sums for items of work.

The Red Book is administered by the engineer (a third party) rather than the contractor or the employer. The engineer will also be responsible for monitoring the construction work (but still keeping the employer fully informed so that he/she may make variations) and certifying payments.

The contents of the book are as follows:

- General Conditions
- Guidance for the Preparation of the Particular Conditions
- Forms of Tender and Contract Agreement
- Dispute Adjudication Agreement

The Red Book provides conditions of contract for construction work where the design is carried out by the employer. The current Red Book bears little resemblance to its predecessors. Earlier versions of the Red Book were drafted for use on civil engineering projects. The current edition drops the words "civil engineering" from the title, and this signifies a move away from the Red Book only being applicable to civil engineering works.

6.3.1.3 The Pink Book—'MDB Harmonized Edition'

This is a variant of the Red Book. It is drafted for use on projects that are funded by certain Multilateral Development Banks (MDBs), which are supranational institutions such as the World Bank, where the Red Book would otherwise have been applicable. Where the project is funded by an MDB but the employer is not responsible for the design, the parties should not use the Pink Book, but use and amend one of the other FIDIC forms of contract.

Prior to the publication of the Red Book, where the MDBs had originally adopted the Red Book for the projects they were funding, they amended the FIDIC General Conditions. As a result of negotiations between FIDIC and the MDBs, the Pink Book was drafted and incorporated the amendments commonly inserted by the MDBs.

The contents of the book are as follows:

- General Conditions
- Guidance for the Preparation of the Particular Conditions
- Forms of Tender and Contract Agreement
- Dispute Adjudication Agreement

As part of their standard bidding documents, the MDBs have, for a number of years, required their borrowers or aid recipients to adopt the FIDIC Conditions of Contract.

The FIDIC MDB edition of the Red Book simplifies the use of the FIDIC contract for the MDBs, their borrowers, and others involved with project procurement, such as consulting engineers, contractors, and contract lawyers.

The following MDBs have all participated in the preparation of this edition of the Red Book:

- African Development Bank
- Asian Development Bank
- Black Sea Trade and Development Bank
- Caribbean Development Bank
- European Bank for Reconstruction and Development
- Inter-American Development Bank
- International Bank for Reconstruction and Development (the World Bank)
- Islamic Bank for Development Bank
- Nordic Development Fund

6.3.1.4 The Yellow Book—'Conditions of Contract for Plant and Design-Build for Electrical and Mechanical Plant and for Building and Engineering Works, Designed by the Contractor' (the Plant and Design-Build Contract)

This form of contract is drafted for use on projects where the contractor carries out the majority of the design (i.e., the contractor carries out the detailed design of the project so that it meets the outline or performance specification prepared by the employer). The Yellow Book is, therefore, traditionally used for the provision of plants and for building or engineering work on a design-build basis.

The Yellow Book is a lump sum price contract with payments made according to achieved milestones on the basis of certification by the engineer (like the Red Book, the engineer administers the contract). The contractor is also subject to a fitness-for-purpose obligation with respect to the completed project.

The contents of the book are as follows:

- General Conditions
- Guidance for the Preparation of the Particular Conditions
- Forms of Tender and Contract Agreement
- Dispute Adjudication Agreement

The Yellow Book provides conditions of contract for construction work where the design is carried out by the contractor. The current Yellow Book bears little resemblance to its predecessors. The current edition drops the words "electrical and mechanical works" from the title and in line with the rest of the FIDIC suite, the focus is now more on the type of procurement rather than the nature of the works.

6.3.1.5 The Orange Book—'Conditions of Contract for Design-Build and Turnkey'

The contents of the book are as follows:

- General Conditions
- Guidance for the Preparation of the Particular Conditions
- Forms of Tender and Agreement

The Orange book was published in 1995 to provide a design and build option to the then-current FIDIC suite. The Orange Book was the first FIDIC contract to adopt the now-current FIDIC style of drafting and was a template for the drafting teams when preparing the 1999 suite of contracts.

6.3.1.6 The Silver Book—'Conditions of Contract for EPC/Turnkey Projects' (the EPC Turnkey Contract)

The Silver Book is drafted for use on EPC (engineer, procure, and construct) projects. These are projects that require the contractor to provide a completed facility to the employer that is ready to be operated at "the turn of a key." These contracts therefore place the overall responsibility for the design and construction of the project on the contractor.

The Silver Book is used where the certainty of price and completion date is important. The Silver Book allows the employer to have greater certainty as to a project's cost as the contractor assumes greater time and cost risks than under the Yellow Book.

The Silver Book may also be used for privately financed BOT (Build, Operate, Transfer) projects. These are projects where the employer takes total responsibility for the design, construction, maintenance, and operation of a project and wishes to pass the responsibility of construction to the contractor. There is no engineer under the Silver Book as his/her responsibilities are assumed by the employer. Similar to the Yellow Book, the contractor is also subject to a fitness-for-purpose obligation in respect of the completed project.

The contents of the book are as follows:

- General Conditions
- Guidance for the Preparation of the Particular Conditions
- Forms of Tender and Contract Agreement
- Dispute Adjudication Agreement

The Silver Book is suitable for use on process, power, and private infrastructure projects where a Contractor is to take on full responsibility for the design and execution of a project. Risks for completion within time, cost, and quality are transferred to the Contractor and so the Silver Book is only suitable for use with experienced contractors familiar with sophisticated risk management techniques.

6.3.1.7 The Gold Book—'Conditions of Contract for Design, Build and Operate Projects' (the Design Build Operate Contract)

As per FIDIC, the Gold Book is drafted to minimize the risk of rapid deterioration after the handover of a project due to poor design, workmanship, or materials. The Gold Book is therefore suitable where long-term operation and maintenance commitment is required along with design and build obligations. The contractor must operate and maintain the completed project on behalf of the employer for a period of typically 20 years from the date of the commissioning certificate, which is issued at the completion of the construction of the project. During this 20-year period, the contractor must meet certain targets and at the end of this period, the project must be returned to the employer in an agreed condition.

Throughout the 20-year period, the employer owns the plant but the contractor operates it at the contractor's own risk.

However, the contractor has no responsibility for financing the project or ensuring its long-term success.

The contents of the book are as follows:

- General Conditions
- Particular Conditions
- Sample Forms

6.3.1.8 The Blue Book—'Form of Contract for Dredging and Reclamation Works'

As per FIDIC, the Blue Book is suitable for all types of reclamation and dredging work as well as ancillary construction works.

Typically, under such a form of contract, the contractor constructs the work in accordance with the employer's design. However, this form of contract can also be adapted for contracts that include or consist entirely of contractor-designed work.

6.3.1.9 The White Book—'FIDIC Client/Consultant Model Services Agreement'

This form of contract is not for the provision of construction and engineering work despite it also being referred to by the color of its cover. This form of contract is used to appoint consultants to provide services to the employer such as feasibility studies, design, contract administration, and project management. The White Book typically forms the basis of the agreement between the consultant and the employer where the construction and engineering work are undertaken pursuant to a FIDIC contract.

6.3.1.10 The Conditions of Subcontract for Construction—'Conditions of Subcontract for Construction for Building and Engineering Works Designed By the Employer'

Unlike the aforementioned forms of FIDIC contract, no color is given to this form of contract. This is the FIDIC-recommended form of subcontract for use with the Red, Yellow, Silver, and Gold books referred to previously.

6.3.2 Scope of General Conditions of Contract and Attachments

The first part consists of 72 clauses and 160 sub-clauses under the title "General Administrative Conditions of Contract." In this section, the fundamental principles of how to regulate a construction contract are determined.

6.3.3 Scope of Part Two "Special Administrative Conditions of Contract"

The second part explains how employers can make regulations according to required local conditions and work in each separate construction contract, referring to the principles in the first part under the title "Special Administrative Conditions of Contract and Provisioning Principles."

FIDIC General Administrative Conditions of Contracts (Part One) and the Special Administrative Conditions of Contracts (Part Two) have an integrity that covers parties' rights and duties. Therefore, a standard form is not intended for the second part, because the second part has to be prepared separately for each tender. In this context, the FIDIC *Manual of the Special Administrative Conditions of Contracts* is published, providing options for a variety of clauses to help in this regard.

6.4 MASTERFORMAT®

MasterFormat® is the specifications-writing standard for most commercial building design and construction projects in North America. The standard is the most widely used standard for organizing specifications and other written information for commercial and institutional building projects in the U.S. and Canada. It lists titles and section numbers for organizing data about construction requirements, products, and activities. By standardizing such information, MasterFormat® facilitates communication among architects, specifiers, contractors, and suppliers, which helps them meet building owners' requirements, timelines, and budgets. The 16 divisions were first published in 1963, which became Master Format® in 1978. Fifty divisions were introduced in 2004, MasterFormat® titles and numbers are organized into two groups. The 50 divisions of MasterFormat® consist of two groups: the Procurement and Contracting Requirements Group and the Specifications Group. The Procurement and Contracting Requirements Group includes only one division, Division 00 "Procurement and Contracting Requirements." The Specifications Group includes 49 divisions parsed into five subgroups as defined by the Construction Specifications Institute (CSI)'s MasterFormat®. TABLE 6.1 illustrates MasterFormat® titles and numbers as of 2022.

6.5 NEC

In September 1985, the Institution of Civil Engineers (ICE) asked its legal affairs committee to review contract forms used at that time. Due to the problems and issues stated in the previous section, the committee recommended the development of a new family of contracts. the first step was to draft key principles of the contract and its functions (Perry, 2001). This work was done under the supervision of Dr. Martin Barnes, a leading construction project manager, founder and chairman of the UK's Association of Project Management and fellow member of the ICE and, with the assistance of Professor John Perry, at the time Head of Civil Engineering Department at University of Birmingham.

TABLE 6.1

MasterFormat® (2020 ©2020)

Division Numbers and Titles

PROCUREMENT AND CONTRACTING REQUIREMENTS GROUP

Division 00 Procurement and Contracting Requirements

SPECIFICATIONS GROUP

General Requirements Subgroup

Division 01 General Requirements

Facility Construction Subgroup

Division 02 Existing Conditions
Division 03 Concrete
Division 04 Masonry
Division 05 Metals
Division 06 Wood, Plastics, and Composites
Division 07 Thermal and Moisture Protection
Division 08 Openings
Division 09 Finishes
Division 10 Specialties
Division 11 Equipment
Division 12 Furnishings
Division 13 Special Construction
Division 14 Conveying Equipment
Division 15 Reserved
Division 16 Reserved
Division 17 Reserved
Division 18 Reserved
Division 19 Reserved

Facility Services Subgroup

Division 20 Reserved
Division 21 Fire Suppression
Division 22 Plumbing
Division 23 Heating, Ventilation, and Air Conditioning (HVAC)
Division 24 Reserved
Division 25 Integrated Automation
Division 26 Electrical
Division 27 Communications
Division 28 Electronic Safety and Security
Division 29 Reserved

Site and Infrastructure Subgroup

Division 30 Reserved
Division 31 Earthwork
Division 32 Exterior Improvements
Division 33 Utilities
Division 34 Transportation
Division 35 Waterway and Marine Construction
Division 36 Reserved
Division 37 Reserved
Division 38 Reserved
Division 39 Reserved

Process Equipment Subgroup

Division 40 Process Interconnections
Division 41 Material Processing and Handling Equipment
Division 42 Process Heating Cooling, and Drying Equipment
Division 43 Process Gas and Liquid Handling, Purification, and Storage Equipment
Division 44 Pollution and Waste Control Equipment
Division 45 Industry-Specific Manufacturing Equipment
Division 46 Water and Wastewater Equipment
Division 47 Reserved
Division 48 Electrical Power Generation
Division 49 Reserved

Source: The Construction Specifications Institute and Construction Specifications Canada Reprinted with permission from CSI.

6.5.1 First Edition

The work of specifying, designing, and drafting the new contract was sponsored by the Institution of Civil Engineers. It was carried out by a small team of people drawn from different backgrounds and with expert legal contributions. The result was the New Engineering Contract (NEC). After full consultation and trials, it was first published for general use in 1993. Since then, it has been taken up widely by the industry and its clients, and achieved considerable success (Barnes, 2004).

6.5.2 Second Edition

In July 1994, Sir Michael Latham published a report called "Constructing the Team" known as the Latham Report. In this report, Latham identified thirteen principles that should be included in a contract to be effective in modern conditions. These principles are listed:

- A duty of all stakeholders to act fairly.
- Duties of teamwork to achieve "win-win" solutions to problems.
- An interrelated package of documents.
- A comprehensive language and guidance notes included.
- The separation of roles, especially the roles related to the project manager and the adjudicator.
- A choice of risk allocation.
- Variations priced in advance.
- Activity schedules or payment schedules.
- Clear payment periods including interests if there are delays.
- Sure, trust fund for payment.
- Quick adjudication of disputes.
- Incentives for a high level of performance.
- Mobilization payments.

In response to these recommendations together with feedback from early users, a second edition of the NEC was published in November 1995 and the NEC was renamed "The Engineering and Construction Contract."

6.5.3 Third Edition

In July 2005, the third edition of the contract was issued. "The Engineering and Construction Contract" has been widely revised and updated. As a result, all other contracts have been updated as well so that NEC3 with its integrated set of documents remains consistent. Two entirely new contracts were introduced. These are the "Term Service Contract" and "The Framework Contract" (NEC one-stop shop, 2007b as cited in Genessay, 2007) joined in 2010 by a supply contract.

6.5.4 Latest April 2013 Edition

The suite was updated and enlarged to 39 documents in April 2013, including a professional services short contract—the Association for Project Management's standard form for appointing project managers—and an enhanced set of guidance documents. (https://www.neccontract.com/About-NEC/History)

6.5.5 The NEC Family of Contracts

The NEC is the only standard contract that is based on modern best practices in project management. It stimulates forward-looking control of cost, time, and performance. Uniquely, and for the first time, it provides for effective management and control of risk. Much of the NEC is "general purpose" and can be used in all situations. The flexibility of the NEC is achieved by choosing the right contract from the NEC family of contracts and choosing the right options for clauses within the contract. The NEC family is made up of seven standard contracts. They all use the same basic set of procedures, the same management culture, and the same terminology (Barnes, 2004).

NEC family of contracts includes the following (NEC3, 2005):

- NEC3 Engineering and Construction Contract (ECC, June 2005)
- NEC3 Engineering and Construction Subcontract (ECS, June 2005)
- NEC3 Professional Services Contract (PSC, June 2005)
- NEC3 Adjudicator's Contract (AC, June 2005)
- NEC3 Engineering and Construction Short Contract (ECSC, June 2005)
- NEC3 Engineering and Construction Short Subcontract (ECSS, June 2005)
- NEC3 Term Service Contract (TSC, June 2005)
- NEC3 Framework Contract (FC, June 2005)

All types of contracts are usually completed with guidance notes and flowcharts, following the recommendations of the Latham Report. This family is growing, and other NEC forms are constantly in preparation or under consideration (Perry, 2001). Within the NEC, all contract documents share the same format so that the user can get more rapidly familiar with its content (ECC3 Guidance Notes, 2005). Thus, each contract is divided into three main sections: nine core clauses that are identical for each contract, main and secondary options that are different for each contract.

6.5.6 The Nine Core Clauses

The nine core clauses include the following (ECC3 Guidance Notes):

1. **General**—The general core clauses deal with definitions, interpretation, ambiguities, and general introductory matters.

2. **The Contractor's main responsibilities**—This section focuses on the contractor's responsibilities, the provision of work and the design, requirements concerning the equipment and the subcontracting process.
3. **Time**—This section looks at one of the key characteristics of NEC, which is the accepted program.
4. **Testing and Defects**—This part covers important issues such as testing and inspection, correction of failures, and acceptance of defects.
5. **Payment**—This section details the assessment of payments due to the contractor by the project manager.
6. **Compensation Events**—This section defines compensation events and procedures to manage them.
7. **Title**—This **section focuses on the employer's entitlement to plant and materials, together with the removal of equipment and materials within the site.**
8. **Risks and Insurance**—This section deals with the employer's and contractor's risks as well as the insurance requirements for both parties.
9. **Termination**—This section deals with reasons and procedures for termination including the payments involved.

As far as dispute resolution is concerned, following the recommendations of the Latham Report (1994), the NEC requires all disputes to be referred to an independent adjudicator prior to arbitration or litigation.

6.5.7 The Structure of ECC3 Edition

The structure of the Engineering and Construction Contract (ECC) is given:

6.5.7.1 Core Documents

The ECC is published in ten volumes (Genessay, 2007). The core documents are listed:

- NEC3 Engineering and Construction Contract (ECC)
- NEC3 Engineering and Construction Contract Guidance Notes
- NEC3 Engineering and Construction Contract Flowcharts
- NEC3 Engineering and Construction Subcontract

6.5.7.2 Main Option Clauses

The ECC has six main options based on different payment mechanisms, which offer different basic allocations of risk between the employer and the contractor and reflect modern procurement practice. All the main options can be used with the boundary between design by the employer and design by the contractor to set the chosen strategy (ECC3 Guidance Notes, 2005). The six main options are listed:

- Option A (Priced Contract with activity schedule)
- Option B (Priced Contract with bill of quantities)

- Option C (Target Contract with activity schedule)
- Option D (Target Contract with bill of quantities)
- Option E (Cost-Reimbursable Contract)
- Option F (Management Contract)

6.5.7.2.1 *Option A (Priced Contract with Activity Schedule)*

An activity schedule is a list of activities prepared by the contractor that he/she expects to carry out in providing the work. When it has been priced by the contractor, the lump sum for each activity is the price to be paid by the employer for that activity. The total of these prices is the contractor's price for providing the whole of the work, including all matters that are at the contractor's risk. Option A provides staged payments.

6.5.7.2.2 *Option B (Priced Contract with Bill of Quantities)*

A bill of quantities comprises a list of work items and quantities. It is prepared by the contractor for the employer. Standard methods of measurement are published which state the items to be included and how the quantities are to be measured and calculated.

Tenderers price the items, taking into account the information in the tender documents and including all matters that are at the contractor's risk. The employer pays for work done on the basis of actual measurement of those items with quantities.

6.5.7.2.3 *Options C and D: Target Contracts (with Activity Schedule or Bill of Quantities)*

Target contracts are sometimes used where the extent of work to be done is not fully defined or where anticipated risks are greater. The financial risk is shared between the employer and the contractor in the following way:

- The contractor tenders a target price in the form of the prices using either the activity schedule or a bill of quantities. The target price includes the contractor's estimate of defined cost plus other costs, overheads, and profit to be covered by his/her fee.
- The contractor tenders his/her fee in terms of fee percentages to be applied to defined cost.
- During the course of the contract, the contractor is paid the defined cost plus the fee. This is defined as the Price for Work Done to Date (PWDD)

The prices are adjusted for the effects of compensation events, and for inflation if Option X1 is used; for Option D the prices are also adjusted as the work completed by the contractor is measured.

At the end of the contract, the contractor is paid (or pays) his/her share of the difference between the final total of the prices and the final PWDD according to a formula stated in the contract data. If the final PWDD is greater than the final total of the prices, the contractor pays his/her share of the difference. The contractor's share is paid provisionally at completion and is corrected in the final account.

6.5.7.2.4 *Option E (Cost-Reimbursable Contract)*

A cost-reimbursable contract should be used when the definition of the work to be done is inadequate even as a basis for a target price and yet an early start to construction is required. In such circumstances, the contractor cannot be expected to take cost risks other than those which entail control of his/her employees and other resources. He/she carries minimum risk and is paid defined cost plus his/her tendered fee, subject only to a small number of constraints designed to motivate efficient working.

6.5.7.2.5 *Option F (Management Contract)*

The conditions of contracts applied to management contracts are still evolving. In practice, several different approaches are used, for example, scope of services, time of appointment, and methods of fee payment. The terms under which subcontractors are employed are also changing. The ECC management contract is based on the following framework.

The contractor's responsibilities for construction work are the same as those of a contractor working under one of the other options. However, he/she does only a limited amount of construction work himself as stated in the contract data. The project manager has no authority to instruct the contractor to carry out further construction work beyond that stated in the contract data. Any increase in the extent of the construction work to be carried out by the contractor must be the subject of negotiation between the employer and the contractor.

The contractor's services apply mainly to the construction phase, although he/she would usually be appointed before construction starts. If substantial pre-construction services are required and the employer wishes to have the option to change the management contractor before construction starts, a separate contract should be awarded for such pre-construction services using the NEC Professional Services Contract.

All subcontracts are direct contracts with the contractor, who acts as a management contractor. If the employer wishes to be a party to the construction subcontracts, a management contract is not appropriate. He/she should instead appoint a construction manager as the project manager and use the ECC with appropriate main options for the contracts with package contractors.

The contractor tenders his/her fee and his/her estimated total of the prices of the subcontracts. The subcontract prices are paid to the contractor as part of the defined cost. The contractor is responsible for supplying management services, including the management of design, if required. If the contractor wishes to be responsible for doing work other than management contracts, he/she must state the extent of that work in the contract data.

The contractor's fee will increase if the subcontractors' prices (part of the defined cost to the contractor) increase due to compensation events. However, he/she will not receive separate payment for his/her work in dealing with compensation events and he/she will not receive any additional fee for work on compensation events which does not lead to an increase in subcontractors' prices.

6.5.7.2.6 *Secondary Option Clauses*

Once the employer has selected the main option, he/she can select the available secondary options to refine the contract strategy (NEC one-stop shop, 2007 cited in

Genessay, 2007). The chosen main option clause and the secondary clauses have to appear on the first statement to be a part of the contract data (ECC3 Guidance Notes, 2005). The secondary options are listed:

- Option X1—Price adjustment for inflation.
- Option X2—Changes in the law.
- Option X3—Multiple currencies.
- Option X4—Parent company guarantee.
- Option X5—Sectional option.
- Option X6—Bonus for early completion.
- Option X7—Delay damages.
- Option X12—Partnering.
- Option X13—Performance bond.
- Option X14—Advance payment to contractor.
- Option X15—Limitation of contractor's liability.
- Option X16—Retention.
- Option X17—Low performance damages.
- Option X18—Limitation of liability.
- Option X20—Key performance indicator.
- Option Y(UK)2—Housing Grants Act.
- Option Y(UK)3—Rights of Third Parties Act.
- Option Z—Additional conditions.

6.5.7.2.7 Dispute Resolution Options

There are two options for dispute resolution under ECC3 which are as follows:

1. **Option W1**—Not **HGCRA** (Housing Grants, Construction and Regeneration Act, 1996) compliant.
2. **Option W2**—For **UK Construction Contracts with HGCRA compliance.**

6.5.4 NEC April 2013 Edition

The latest edition of the NEC3 contains 39 documents including the new Professional Services Short Contract which has been developed in partnership with the Association for Project Management and a series of seven new "how to" books designed to aid the use of these contracts. Also included are the latest Construction Act amendments, and Project Bank Account provisions and clauses as well as CIC BIM protocol references.

6.5.4.1 NEC Suite of Contracts, April 2013 Edition

1. NEC3 Engineering and Construction Contract (ECC)
2. NEC3 Engineering and Construction Contract Option A: Priced contract with activity schedule
3. NEC3 Engineering and Construction Contract Option B: Priced contract with bill of quantities

4. NEC3 Engineering and Construction Contract Option C: Target contract with activity schedule
5. NEC3 Engineering and Construction Contract Option D: Target contract with bill of quantities
6. NEC3 Engineering and Construction Contract Option E: Cost reimbursable contract
7. NEC3 Engineering and Construction Contract Option F: Management contract
8. NEC3 Engineering and Construction Subcontract (ECS)
9. NEC3 Engineering and Construction Short Contract (ECSC)
10. NEC3 Engineering and Construction Short Subcontract (ECSS)
11. NEC3 Professional Services Contract (PSC)
12. NEC3 Professional Services Short Contract (PSSC)
13. NEC3 Term Service Contract (TSC)
14. NEC3 Term Service Short Contract (TSSC)
15. NEC3 Supply Contract (SC)
16. NEC3 Supply Short Contract (SSC)
17. NEC3 Framework Contract (FC)
18. NEC3 Adjudicator's Contract (AC)

6.5.4.2 Guidance Notes and Flow Charts

1. NEC3 Engineering and Construction Contract Guidance Notes
2. NEC3 Engineering and Construction Contract Flow Charts
3. NEC3 Engineering and Construction Short Contract Guidance Notes and Flow Charts
4. NEC3 Professional Services Contract Guidance Notes and Flow Charts
5. NEC3 Term Service Contract Guidance Notes
6. NEC3 Term Service Contract Flow Charts
7. NEC3 Term Service Short Contract Guidance Notes and Flow Charts
8. NEC3 Supply Contract Guidance Notes
9. NEC3 Supply Contract Flow Charts
10. NEC3 Supply Short Contract Guidance Notes and Flow Charts
11. NEC3 Framework Contract Guidance Notes and Flow Chart
12. NEC3 Adjudicator's Contract Guidance Notes and Flow Charts
13. NEC3 Procurement and Contract Strategies
14. NEC3: How to Write the ECC Works Information
15. NEC3: How to Use the ECC Communication Forms
16. NEC3: How to Write the PSC Scope
17. NEC3: How to Use the PSC Communication Forms
18. NEC3: How to Write the TSC Service Information
19. NEC3: How to Use the TSC Communication Forms
20. NEC3: How to Use BIM with NEC3 Contracts

6.5.5 Implementation of the NEC

6.5.5.1 Bodies Responsible for NEC's Implementation

This section deals with the bodies responsible for the implementation of NEC in the UK and overseas. From the literature, four major bodies are listed (Genessay, 2007):

1. Institution of Civil Engineers—As a qualifying body, a network for sharing knowledge, a supplier of resources, the worldwide recognized ICE is an ideal support to the implementation of NEC that it launched back in the the 90's. Through the NEC, ICE has been encouraging innovation and excellence in the profession (http://www.ice.org.uk/About-ICE, 2010).
2. Thomas Telford Limited—Thomas Telford Ltd is the business subsidiary of ICE. It is the administrative support for the NEC through its NEC division which manages the training, talks, workshops, seminars, publications, and sales of the NEC family of contracts (http://www.thomastelford.com/, 2010).
3. NEC User's Group—The NEC User's Group is made up of more than two hundred member organizations representing clients, consultants, designers, project managers, and contractors. Its aim is to stimulate members to stay actively involved in the ongoing development and promotion of the ECC by sharing ideas and experiences (http://www.neccontract.com/users_group/index.asp, 2010).
4. NEC panel—The NEC panel has replaced the earliest NEC Working Group. Composed of senior professionals of the construction industry who meet on a regular basis, the NEC panel is ICE's official consultative board for the NEC (http://www.neccontract.com/news/article.asp?NEWS_ID=571, 2006).

7 ISO Certification in Construction Industry

Mustafa Shraim

7.1 INTRODUCTION

In the construction sector today, quality workmanship provides a competitive advantage in terms of customer satisfaction and loyalty. Quality organizations provide a safe working environment for their crews and a safe living environment for future residents. Although construction site accidents are not as common in developed countries, even one accident is one too many. The loss of human life, property, credibility, and resources is often the result of poor quality control.

Quality in construction management has a lengthy history comparable to manufacturing, but with its own unique characteristics (Arditi & Gunaydin, 1997):

- Almost all construction projects are different from each other and constructed in varying conditions.
- The lifecycle of a construction project is longer than the lifecycle of most manufactured products.
- There is no uniform standard used to evaluate overall construction quality as in manufacturing, therefore, evaluations of construction projects tend to be more subjective.
- Various participants in construction—project-owner, designer, general contractor, subcontractor, materials supplier—have different ideas of what quality workmanship is.

Because of these distinguishing characteristics, the construction industry has not been able to implement quality procedures as effectively as the manufacturing sector. Defect prevention, however, is desirable in both sectors. The rule of quality cost is called "1-10-100," meaning it only costs one unit to detect a quality issue in the design phase. It will cost ten times more if the defect is identified during the installation stage and 100 times more if the defect is discovered after the product is installed or purchased by the customer (Sowards, 2013).

Another problem is that the costs of poor construction are not really known because defects are often not recorded. In order to reduce defects, construction managers need to implement a system to better manage their processes. One way to do this is to adopt a quality framework, such as the International Organization for Standardization (ISO) framework for planning and design, load computations, operations, and maintenance.

 DOI: 10.1201/9781003636687-7

7.2 IMPORTANCE OF STANDARDS

Quality measures can be embedded in various philosophies, management systems, tools, and techniques. ISO certification can help construction companies with customer or regulatory compliance, gain a competitive advantage, improve operational efficiency, and send a credible signal of quality to current and potential clients (Kale & Arditi, 2006). Although ISO does not define product quality, it ensures that customer feedback is captured and measured to improve future processes (Ollila, 2011).

Large construction companies have taken advantage of ISO certification whereas small and medium-sized firms have not because of fewer resources such as staff, money, and time. "We have a lot of skilled engineers, engineers . . . [who] like to focus on engineering and are somewhat resistant to documentation and procedures," stated a vice president of a manufacturing/design firm (Libby, 2007, p. 102). Therefore, many cost overruns, reworks, delays, and disputes can be traced back to inferior design, poor contract administration, and lax supervision by the client (Chini & Valdez, 2003; Tan & Lu, 1995).

7.3 STANDARDS ORGANIZATIONS

Many organizations want to systematize the implementation of business management systems in terms of different functions and operations such as quality management (ISO 9000, IATF 16949, ISO 13485, etc.), environmental impact (ISO 14000 and EMAS) and corporate social responsibility (SA 8000, AA 1000, and ISO 26000). When comparing individual countries' standards to ISO, ISO appears to be the common denominator for most standardization attempts. For instance, the ASN of Senegal, SAC of China, ELOT of Greece, BSI of the UK, BIS of India, ABNT of Brazil, SCC of Canada, and ANSI of the USA, all have ISO in common.

7.4 INTERNATIONAL STANDARDS ORGANIZATION (ISO)

To enhance efficiency, competitiveness, and customer satisfaction, an increasing number of organizations are adopting the philosophy of total quality management (TQM). In many instances, this philosophy is operationalized through the implementation of a quality management system (QMS) that uses the International Organization for Standardization (ISO) 9000 series of standards.

The ISO, founded in 1947, is headquartered in Geneva, Switzerland. ISO's purpose is to aid worldwide commerce in implementing international quality standards for products and services (Stevenson & Barnes, 2001). The underlying premise of ISO 9000 certification is the creation of products and services that utilize a system of inputs and outputs that add and measure value throughout the manufacturing process.

Unlike TQM which originated in Japan, ISO certification began in Europe in 1987 and quickly spread to North America, Japan, and the rest of the world. ISO certification is a quality management program that encourages companies to shift their mindsets from the traditional reactive approach to a proactive management approach (Kale & Arditi, 2006). The ISO's certification process provides worldwide standards to support external quality assurance processes and to aid in global commerce. Having met the certification requirement, partners, buyers, and customers gain assurance that a company's products meet certain standards of quality.

The ISO series consists of two basic types of standards: I) those addressing quality assurance and II) those addressing quality management (Arditi & Gunaydin, 1997). ISO 9001 is designed for contractual and assessment purposes while ISO 9004 provides guidance for companies developing and implementing quality management systems.

To receive ISO 9001 certification, an accredited third party (an approved external auditor) evaluates the company's processes and documentation and determines whether the organization is systematically following the policies and procedures necessary to produce high-quality products. Companies using QMS will generally experience an improvement in the quality of products and services, a reduction in quality-related costs, more efficient internal business processes, and the creation of corporate memory through proper documentation and retention (Vrassidas & Chatzistelios, 2014).

According to Renuka and Venkateshwara (2006), the benefits of ISO certification are:

- Higher productivity and sales,
- Better control of business, reduced costs, and fewer customer complaints,
- Improved quality of work life,
- Increased customer preference,
- Improved company image, customer relationships, and competitiveness in the marketplace,
- Streamlined procedures and documentation, and
- Increased consciousness for preventive and corrective actions.

On the other hand, critics of ISO 9001 argue that it does more harm than good. Seddon (1998) criticizes ISO 9001 as an inspection-oriented tool, not a development-oriented process. Since quality is defined by external auditors, critics believe ISO can make matters worse for customers by not delivering promised results. Others believe certification is too costly, representing a pursuit of certification rather than quality (Brown, 1994). Some view ISO as a paper-driven, overly bureaucratic process (Curkovic & Pagell, 1999; Quazi et al., 2002; Boiral & Amara, 2009; Boiral, 2011).

Critics also argue that inconsistencies are inherent in the evaluation process (Stevenson & Barnes, 2001). For instance, since regulations and implementation of the standards are left up to each participating country, no single set of guidelines exists, and the amount of required work for ISO varies. Furthermore, the cost of ISO implementation is significant, particularly for a small organization with limited employee participation (Martinez-Costa & Martinez-Lorente, 2007a, 2007b).

Since ISO 9000 is not industry-specific, critics feel ISO fails to address the unique problems inherent in some industries (Stevenson & Barnes, 2001). The transportation, chemical, oil, computer, and electronics industries have traditionally had strong ties to ISO 9000. On the other hand, some industries like tool manufacturers, auto manufacturing, and steel believe the ISO 9000 certification does not address their specific needs and have developed their own sets of standards.

Symonds (1998) offers counterarguments in favor of the ISO standard. He believes people who criticize ISO are usually those not directly involved with its implementation and should not blame the tool, but rather poor execution of the tool. Harrington (1997) studied 60 organizations to analyze best practices. He concluded that failure or success depends on management—if management understands how to implement ISO, then it will succeed. Otherwise, no matter what quality tools management uses, improvements will not succeed.

Proponents of ISO 9000 further argue that proper implementation leads the organization to better performance by increasing sales volume, lowering customer complaints, reducing the variance in the production process, and increasing competitive advantage.

Despite the discussion for and against certification, the number of certifications around the world is increasing. As of the end of 2023, the International Organization for Standardization has a membership of 170 national standard bodies (countries). The number of ISO 9001 QMS certifications in developing countries is also increasing. In fact, the number of certifications in developed countries has stabilized or even decreased whereas certifications in developing countries such as China, India, and some countries in Africa, have increased. As reported in the ISO Survey 2022, the top ten countries with the highest number of ISO certifications are listed in TABLE 7.1. (International Organization for Standardization (2022)—The ISO survey of management system standard certification.)

TABLE 7.1
Top Ten Countries with ISO Certification

Serial Number	Country	Number of Certifications
1	China	551,855
2	Italy	94,216
3	India	61,653
4	Germany	47,576
5	United Kingdom	43,765
6	Japan	38,916
7	Spain	32,059
8	United States of America	29,579
9	Korea (Republic of)	27,155
10	France	21,880

Note: *(Source: The ISO Survey 2022, at iso.org)

7.5 ISO 9001 IMPLEMENTATION PROCESS

According to Kale and Arditi (2006), the ISO implementation process follows three stages:

1. Mapping or documenting routines that underlie an organization's internal and external relationships,
2. Improving internal and external relationships by making changes and setting up essential business processes in order to bring customers and suppliers together, to achieve continuous flow of accurate and reliable information from all parties to reduce guesswork and to increase speed and flexibility, and
3. Adherence to improved internal and external organizational routines.

Unlike the manufacturing sector, the construction industry has its unique challenges. Construction companies provide services in ever-changing locations, deal with different stakeholders, and lack fringe benefits for employees, making it more difficult to incorporate standardization. Thus, the very nature of this industry makes systemization, standardization, and process automation difficult. When implementing a QMS, the industry struggles with: (Cachadinha, 2009)

- Systematization and structuring,
- Document control,
- Defining and maintaining procedures,
- Client satisfaction including evaluation, analysis, and action,
- Interaction between quality and production departments; and
- Cost and assignment of human resources.

During the first half of the twentieth century, engineers and architects controlled the design phase of construction (Arditi & Gunaydin, 1997). They ensured clients (owners) received quality construction. In the 1950s and 1960s, clients became increasingly concerned with costs, but still with emphasis on quality. At about the same time, however, the practice of sealed competitive bidding forced many general contractors to cut costs in order to maintain profitability. As mechanical and electrical systems became more complex, the general contractor turned responsibility for such work over to subcontractors, giving up general oversight regarding the quality of workmanship.

Due to increased complexity and stricter regulations in the 1980s and 1990s, construction management firms began to perform managerial functions on behalf of the client from inception to the completion of the construction project. Consequently, inspections once performed by architects and engineers are now done by construction management firms.

The following case is about a construction management firm that developed and applied a QMS that was originally certified to ISO 9001:2008 in 2013 and then transitioned to ISO 9001:2015 (ANSI/ISO/ASQ Q9001:2015. *Quality Management Systems—Requirements*).

7.6 CASE STUDY—ISO IMPLEMENTATION AT RESOURCE INTERNATIONAL INC. (RII)

This case study is based on direct participation of the authors as ISO consultants for Rii. Incorporated in 1973, Farah Majidzadeh, CEO and Chairperson, founded Rii in the basement of her home in Columbus, Ohio. Rii is a family-run business. Dr. K. Majidzadeh, a retired Ohio State University professor of civil and structural engineering, provides engineering oversight, while his wife, Farah, oversees the business aspects of Rii. All four children of the founder work in top management positions.

The company is based in Columbus, Ohio, USA with additional offices in Cleveland, Cincinnati, and Indianapolis and more than 150 employees. The company is characterized by a strong, centralized hands-on leadership style. Rii offers a full-service consulting, engineering, and construction management firm assisting public and private clients in the following areas:

- Civil Engineering
- Construction Administration, Engineering, and Inspection Services
- Construction Management
- Environmental Services
- Geotechnical and Material Engineering Services
- Ground Penetrating Radar
- Information Technology
- Asset Management and Pavement Engineering
- Survey, Mapping, and SUE (Subsurface Utility Engineering)

Rii took the ISO path for three reasons:

1. To enhance its professional image and assurance of quality,
2. To gain a competitive edge by improving its internal quality systems, and
3. To improve its capability to bid on larger projects, both public and private, especially because many major public works projects are awarded to companies with a certified QMS (Cachadinha, 2009).

7.6.1 Methodology

Mafi and Shraim followed the model (FIGURE 7.1) that was used by the consulting company Rii hired to help design and implement a QMS. Rii management also wanted an ISO 9001 system that minimized the long-term burdens of sustaining it.

7.6.1.1 Design and Implementation of the QMS

Mafi and Shraim met with Rii leadership to devise a plan and timeline for this project. The authors asked the leadership to identify an individual to serve as a change agent who had the authority and credibility to spearhead this initiative and a project coordinator who could oversee the details of the project and serve as the point person for all the questions and documents related to the certification process.

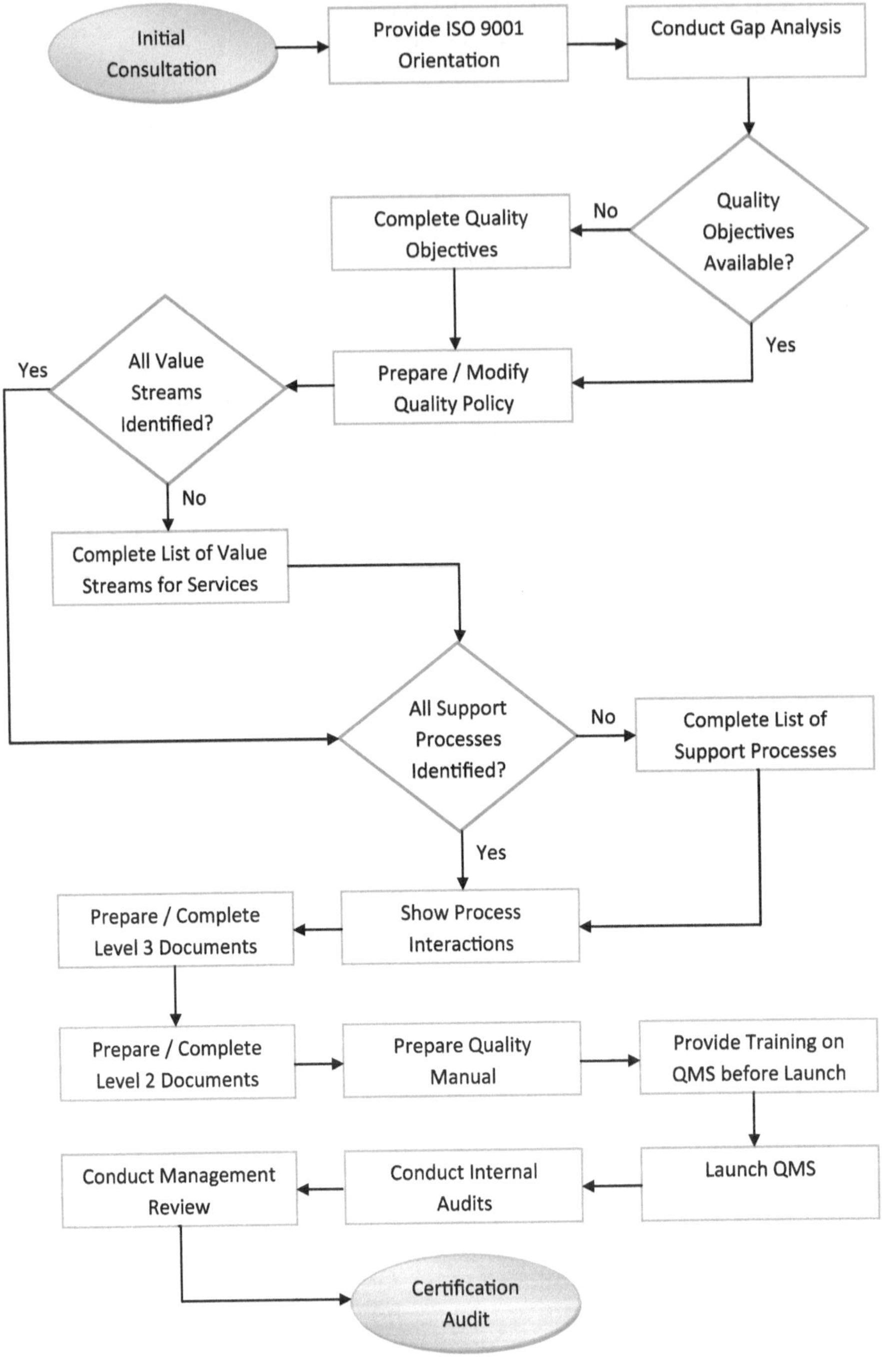

FIGURE 7.1 ISO 9001 implementation model.

The five phases of the QMS development process are pre-implementation, defining and establishing quality objectives, determining the processes (value streams and support processes), documentation (all levels), and the QMS launch and certification.

7.6.1.1.1 Pre-Implementation

At the initial consultation, the authors gathered general information about Rii to determine the scope of the project. Such information included size, locations, number of employees, services provided, and any permissible exclusions. This helped the authors prepare a gap analysis to determine the current extent of compliance to the standards.

At that time, the company was trying to decide whether to apply for certification. Skeptics in the company did not think the company was ready for such a project. Others thought the process would add a significant amount of administrative work to the organization and, therefore, could not be justified.

Interestingly enough, Rii leadership selected the CFO of the company, a CPA, to be the change agent for this initiative. When asked about this decision, Dr. Majidzadeh justified it stating, "If we can get the most skeptical person to buy into this process, then it is easy to convince everyone else." The social media/marketing staff served as the coordinators of the company-wide ISO process.

The CFO was the most critical person pertaining to the success of the initiative. Once he/she met with the consultants and shared his/her concerns, they asked him/her to give the process a chance to succeed. At the beginning, he/she estimated the chance of success to be 10%. As the process progressed, his/her expectations rose to 100%. He/she began to meet the directors of each department at lunch meetings, explaining the purpose of the ISO certification and answering questions. Lunchtime was strategically selected to minimize the time spent on certification and build goodwill since employees appreciate a free lunch.

7.6.1.1.2 Defining and Establishing Quality Objectives and Policy

In defining the quality policy and objectives, it is important to review the specific requirements stated in the ISO 9001 standard. In addition, a quality policy must be written, communicated, and reviewed for suitability. Furthermore, management must commit to complying with the ISO requirements and its continuous improvement process which includes developing a framework for establishing and reviewing quality objectives. This requirement suggests that the quality objectives and policies should be determined before the QMS is established.

As for requirements for quality objectives, they must be measurable and consistent with the policy. These objectives are to be established at relevant functions and levels within the organization. This means that higher-level objectives should be implemented at lower levels with measurable goals. Typically, quality objectives include those related to quality of services as well as timely delivery. Other objectives may include customer service and responsiveness.

Defining a quality policy and objectives begins with listing input generated from employees as well as from clients. Employee input from different operations can lead to better policies and measures. Additionally, customer surveys can be conducted in order to understand what is important to clients.

7.6.1.1.3 Determining Processes

Utilizing the process approach is a critical element when it comes to effective management of operations. As described under Clause 0.2 of ISO 9001, the "adoption of a process approach when developing, implementing and improving the effectiveness of a quality management system, . . . enhance[s] customer satisfaction by meeting customer requirements."

Additionally, Clause 4.1 requires the organization to determine processes needed for effective operations as well as the sequence and interactions among such processes. It should be noted that processes can be either value stream or support. A value stream process includes all actions required to bring about the product or service through the main flows (Rother & Shook, 2003). On the other hand, support processes ensure that value streams are effective and meet or exceed customer requirements.

Separating processes into value stream and support categories is beneficial when applying Lean management techniques. For example, value stream maps can show both current and future states so that improvement can be easily implemented and documented.

7.6.1.1.4 Value Stream Processes

In general, the approach Mafi and Shraim used for designing and implementing a QMS starts with identifying the services being provided. For example, what revenue-generating services does Rii provide to its client base?

To address this question, all department heads should be brought together for a session to report all of the sources of generating revenue for the company. After a list is created, these services are grouped based on commonalities. For example, an IT company might provide services such as website design and development, search engine optimization, network system design and integration, barcode systems, warehouse automation, online support, and staffing of IT professionals. This list might then be grouped into families as follows:

- Website Services: Website design, web hosting, and search engine optimization.
- Warehouse Services: Warehouse automation and barcode systems.
- Network Services: Network system design and integration.
- Expert services: Online support and staffing of IT professionals.

Value stream services were grouped into the following families at Rii:

- Design Engineering
- Environmental Services
- Construction Management
- Surveying Services
- Geotechnical Services
- Construction Inspection and Testing
- Information Technology Services

Each of these families represented services that are managed in a similar manner. For example, the construction, inspection, and testing family includes specific tests conducted in the lab and on-site. What is common among these tests is that they are managed in the same way. Although each family has separate "how to" documents, the overall management process is the same.

7.6.1.1.5 Support Processes

Support processes are divided into two groups:

- Processes/procedures required by the standard.
- Processes not required by the standard but important for sustaining the value streams as well as the quality policy and objectives.

The processes required by the procedures include document and record control, internal audits, control of non-conformance and corrective and preventive actions. Additional processes were developed based on the importance of achieving results influenced by regulations, company policies, and value stream operations. Such processes included:

- Management Review
- Feasibility Review
- Purchasing and Supplier Control
- Training
- Client Feedback Management

After determining all value stream and support processes needed for the QMS, the authors had to determine the relationships and interactions among the processes. TABLE 7.2 illustrates how process relationships, interactions, and performance measurements are documented. The performance of all processes must be linked (directly or indirectly) to quality objectives and ultimately to the performance of Rii.

7.6.1.2 Documentation

In general, quality management systems begin with the writing of a quality manual. The quality manual serves as a roadmap for the QMS. It is more practical to build the "road" before preparing the "map." The manual is generic at first and as the QMS develops, the manual is updated.

A team including the manager and director responsible for each value stream process meets to document the existing value stream processes. The team is provided with a template to document value stream processes in the following format:

- Service Provided
- Responsibility
- Outsourced or Subcontracted
- Input
- Activities

TABLE 7.2
Example of Process Inputs, Activities, Output, and Measurement

Process Name	Input	Process Activities & Reference	Primary Output	Output Measurement & Records
Document and Record Control	>ISO 9001 requirements. >Other standards. >Process requirements. >Legal requirements.	Procedure QSP-1: Outline of activities that convert input into output.	Controlled Documents	>Doc Control Database Reports >Internal and External Audits
Management Review	>Review input.	Procedure QSP-2: Outline of activities that convert input into output.	>Decisions and Actions >Improvement Plans	>Management Review Reports >Action Database Status >Internal and External Audits
Purchasing and Supplier Control	>Quotes >Sub-consultant proposals. >Sub-consultant performance.	Procedure QSP-3: Outline of activities that convert input into output.	>PO Issued >Signed Contracts >Received Items	>Internal/External Audits >Supplier Evaluation

- Output/Record
- Measurement of Performance

After completing this first step, the plans are reviewed to ensure that everything is accounted for including external references, forms, and records. Additionally, process performance measures were verified and modified. If other documents exist that are a part of the value stream process, they are referenced at this time. At Rii, all existing forms and records were utilized with minor adjustments.

There are three levels of documents in the QM Manual.

1. Level 3 documents: This level includes internal work instructions, external testing procedures, and other "how to" standards. Some of the internal documents already existed but did not include controls. Since all external testing procedures, standards, and manuals were already available, the authors were able to:
   - Take a full inventory of all available documents (hardcopies and electronic),
   - Remove redundant documents,
   - Assign responsibilities for creating new documents, and adjusting existing ones, and
   - Document controls.

2. Level 2 documents: This level includes documents that describe value stream and support processes. If internal work instructions or other external Level 3 documents are available for any of these processes, they are referenced at this level. In summary, each of the procedures at this level (value stream or quality system procedures) is considered the parent document for relevant Level 3 documents.
3. Level 1 documents—The Quality Manual: After the Level 2 and 3 documents are prepared, the quality manual is prepared in Level 1. The quality manual for Rii addresses the following:
   - Requirements in the ISO standard,
   - Legal and company requirements,
   - Reference to procedures required by the standard, and
   - A table describing the interactions between QMS processes.

A quality management system must cover all the activities leading to the turn out of the final product or service. The quality system depends entirely on the scope of operation of the organization and particular circumstances such as the number of employees, skill level, type of organization, and physical size of the premises of the organization. Adoption of a quality management system is a strategic decision for the organization. The quality manual is the document that identifies and describes the QMS. The QMS is based on the guidelines for performance improvement as per ISO 9004 and the quality management requirements. ISO 9000 standard outlines the necessary steps to implement the QMS. These are:

1. Identify the processes (activities and necessary elements) needed for the QMS.
2. Determine the sequence and interaction of these processes and how they fit together to accomplish quality goals.
3. Determine how these processes are effectively operated and controlled.
4. Measure, monitor, and analyze these processes and implement action necessary to correct the process and achieve continual requirements.
5. Ensure that all information is available to support the operation and monitoring of the process.
6. Display the most options, thus helping make the right management system.

These documents are regularly updated and revised to meet the changing industry requirements due to:

- Globalization.
- Higher levels of expectations by customers.
- Higher performance requirements.
- Complex nature of environment and work.
- Specialized services requirements.

7.6.1.3 QMS Launch and Certification

Before the QMS is launched, all documents are reviewed for corrections. Applicable copies of the final QMS are made available and accessible at all locations.

Although everyone is aware of their responsibilities, QMS training is provided for all associates. In addition, associates receive additional training on their specific value streams. At this point, the QMS is launched as scheduled. In addition, a date for the certification audit is established with the selected registrar.

Prior to certification, an internal audit is required for the entire QMS. In this case, minor issues were found and fixed. Results of the internal audits, as well as other inputs, were evaluated at the management review session. Now, Rii is ready for the certification audit.

When the audit occurred, Rii associates were well prepared with everyone assuming ownership of the process. The audit team was very impressed with the QMS and its results. The team reported no nonconformities and recommended a few opportunities for improvement. Rii received its certification in the year 2013, the same year the company celebrated its 40th anniversary.

7.6.1.4 Lessons Learned

Even a quarter of a century after the debut of the ISO 9000 standards, the standards have had a worldwide impact on trade and quality. Despite the costs and complexity of the certification process, research shows that the certification process and its maintenance is worth the investment that pays off many times over if done correctly. During an interview, the CFO commented that the ISO process did not bring drastic changes to the Rii culture but has had a significant positive impact on documentation control and maintenance bringing consistency to their documentation processes. Furthermore, ISO helped Rii catch lapses in QMS that would have gone unnoticed in the past (i.e., customer complaints). Now all complaints are tracked and recorded so that improvements can be made.

Most employees, especially the engineers, were reluctant to add yet another "job" to their daily routine. As they experienced how ISO improved their work, their resistance diminished. Rii leadership hopes that the ISO processes will become second nature for employees. ISO certification has also helped improve communications within the organization and external parties thereby improving client relationships. Standardization, consistency, and accountability across the entire organization now exist at Rii.

The authors learned the following valuable lessons on how to better implement a QMS:

- Management buy-in and commitment serve as the cornerstone for completing an ISO project on time. If top management leads the implementation efforts, the entire organization is more likely to participate in the process.
- Unlike manufacturing, value streams in service organizations are more difficult to identify since they are not related to a tangible product. Value stream identification was difficult because Rii provides many different services.
- Managers found brainstorming sessions helpful in outlining their value stream processes. Each brainstorming session addressed services offered, inputs and outputs, and performance measures.
- Rii subcontracted its internal audits to a local consulting firm to ensure the audits were effective in uncovering issues.

- In the beginning, the implementation efforts might meet resistance. Providing basic training on requirements and answering questions at the onset helps to alleviate anxiety and improve participation.

7.6.1.5 Conclusion

Clearly, a primary goal in construction safety is zero on-the-job accidents and zero quality defects. Most construction managers, even if they agree with these standards, still accept defects in construction as commonplace, thinking that it is impossible to deliver a defect-free product. The best target, however, is no defects, no reworks, and no punch lists (Sowards, 2013). This chapter demonstrates the journey Rii took to implement ISO Standards.

Findings of quality management practices are as applicable today as they were when first created nearly 30 years ago (Arditi & Gunaydin, 1997). Top management commitment to quality in each phase of the construction process, education of basics of quality management, statistical methods, and problem-solving tools in monitoring construction progress to architects and engineers, lead to TQM. Facilitation of teamwork among manufacturers, subcontractors, main contractors, vendors, designers, project managers, and clients (owners) is also part of the TQM process.

For success, the requirements of the client (owner) must be clearly defined at the beginning of the project. The more time spent in planning, the more smoothly the project will run. Similarly, drawings and specifications received from designers should be clear, concise, and uniform since they are the only documents a constructor sees regarding the design concept, size, and scope of the job.

No system can replace human knowledge and know-how, but a quality management system can help organizations by documenting human knowledge in a systematic way. Going through the journey of ISO certification is not an easy one but well worth it. However, some organizations, after the initial success, fail to sustain their momentum and fall back into the old habits. Thus, a regular management review helps companies maintain and improve their TQM programs.

7.7 ISO 14001:2015

ISO 14001 provides the requirements for environmental management systems by identifying measurable performance goals that verify improvement. This standard is especially important for the construction sector because this sector builds the infrastructures that house human life and can adversely affect the environment with land deterioration, resource depletion, waste generation, air pollution, noise pollution, and water pollution (Turk, 2009). For example, the annual solid waste generated from construction activities in China is around 30% to 40% of the total municipal waste (Zeng et al., 2003), and the amount of annual construction waste per capita in the EU countries is estimated to be .5 to 1 ton (Ekanayake & Ofori, 2004). It is also estimated that construction activities consume approximately 40% of the world's material and energy flow (Lenssen & Roodman, 1995). In addition, many of the materials are irreplaceable raw materials. These and similar issues attest to the importance of the construction sector's compliance with ISO 14001 requirements.

A study by Turk (2009) found construction companies in Turkey seek certification for the following reasons: easier access to the international markets, per the requests from clients, to beat other competitors, and in anticipation that ISO 14001 standards will be mandatory soon for reasons similar to those described for ISO 9001.

7.8 ISO 27001

ISO 27001 provides the requirements for information security management systems (ISMS). As with ISO 9001 and 14001, if the process is correct, then the outcome will be satisfactory (Gillies, 2011). Organizations can develop and implement a framework for managing the security of their information assets including financial information, intellectual property, and employee details, or information entrusted to them by clients or third parties. These standards prepare companies for an independent assessment of ISMS as applied to the protection of information.

Studies suggest that ISO 27001 adoption has been slower compared with other ISO standards (Gillies, 2011). Costs appear to be a major barrier for most small companies. However, with further globalization of businesses, there is more legitimacy for ISO 27001 certification either by government regulations (i.e., Japan), supplier/buyer demands, or the necessity for outsourcing and off-shoring (Gillies, 2011).

Another important driver for this certification is to provide evidence to partners that the company has identified and measured their security risks, implementing a series of procedures to minimize those risks. A further incentive is lower insurance premiums for ISO 27001 certified companies (Von Solms & Von Solms, 2005).

The more recent requirements have now replaced the earlier 2005 version. Information security has always been a national and international issue, and this new version of the standard reflects eight years of improvements in the understanding of effective information security management. It also takes account of the evolution in the cyber threat landscape over that period and allows for a new range of best practices. Information security is now a management issue and governance responsibility. TABLE 7.3 lists the number of certifications for the major categories as of 2022 as reported in the ISO Survey 2022 (International Organization for Standardization [2022]—The ISO survey of management system standard certification):

TABLE 7.3
Number of Certification by Categories Number of Certifications by Categories*

Standard	Number of Certifications in 2012	Number of Certifications in 2022
ISO 9001	1,096,987	1,265,216
ISO 14001	284,654	529,853
ISO 27001	19,620	71,549

Note: *(Source: The ISO Survey 2022, at iso.org)

7.9 ISO 9001

The ISO 9001:2015 was published in September 2015. It is also the new core guideline for all management system standards (MSS). This revision includes some changes and modifications over the previous revision.

> The first modification is to promote a more consistent look, language, and feel to all the quality management systems. This is an improvement over earlier revisions because this allows organizations that have or seek multiple certifications to use a common language and structure for all their certification processes. Additionally, the word "leadership" is no longer avoided as in previous standards. The previous clause "Management Responsibility" or "top management," has now become "Leadership" where the emphasis is clearly on the demonstration of leadership.
>
> The second change is to incorporate risk-based thinking into the management system by considering the context of the organization. In other words, all processes are not equal for all organizations with some being more critical than others resulting in different levels of risk. Therefore, it is recommended that organizations must fully explore and understand the internal and external conditions under which they function.

The complexity of today's business practices, along with advances in technology, makes the speed of change faster. ISO 9001:2015 requires the organization to "determine external and internal issues that are relevant to its purpose and its strategic direction and that affect its ability to achieve the intended result(s) of its quality management systems" (ISO 9001:2015). In order to do so, organizations must have a clear understanding of their strategic direction. Often organizations' strategies are uncoordinated and misaligned without clear direction. This lack of direction has a direct impact on its QMS, thus aligning strategies is a great input to any QMS (West & Cianfrani, 2014). Earlier versions of DIS (Draft International Standard) did not emphasize this alignment. While the standards mandate such an inquiry, it is the responsibility of the organization to decide on the extent to evaluate the internal and external conditions pertaining to its performance. This is a challenge, although the concept is fairly simple.

A third addition is the use of the terms "interested parties" and "needs and expectations." Previously, only the customers' voices were addressed. The change ensures that voices other than customers are relevant. For instance, interested parties may include regulatory bodies, suppliers, the community, and society at large. This modification does not negate the voice of the customers but rather acknowledges the voices of other stakeholders. It is up to the organization to decide who the stakeholders are and what issues they have that could potentially impact the organization's performance.

The last major modification proposed is the system thinking and process approach. Although the use of a process approach in developing and maintaining QMS is not new, ISO 9001:2015 requires organizations to identify their processes, interactions, and resources needed to operate. These processes need to be controlled, measured, and improved. ISO 9001:2015 also reiterates that people are key to improving processes and system thinking serves as the catalyst to make this happen. It is critical to understand how the entire system works rather than search only for direct cause-effect relationships.

Changes in ISO 9001:2015 are an opportunity to revisit organizational areas that yet need to be improved. An awareness of the upcoming changes in ISO 9001:2015 will enable quality professionals to better prepare for the future.

It should also be noted that early in 2024, the International Organization for Standardization added an amendment for climate change impact on the quality management system It is now a requirement that each organization determine whether climate change is a relevant issue to its operation (ISO 9001:2015/Amd 1:2024). For documentation purposes, this could potentially be a part of the management review process inputs as this may change over time.

TABLE 7.4 lists Clauses of ISO 9001:2015

TABLE 7.5 lists examples of the content of a quality management manual for engineering consultants (design and supervision) and TABLE 7.6 lists examples of the content of a quality management system manual for building construction organization (contractor).

TABLE 7.4
Clause Details of ISO 9001:2015

Clause No.	ISO 9001:2015		
1.0	**Scope**		
	1.1	General.	
2.0	**Not valid**		
3.0	**Included in Primary Standard, ISO 9000**		
4.0	**Context of the Organization/Quality Management System**		
	4.1	Understanding the organization and its context.	
	4.2	Understanding the needs and expectations of both parties.	
	4.3	Determine the scope of quality management system.	
	4.4	Quality management system and its processes.	
5.0	**Leadership**		
	5.1	Leadership and commitment.	
		5.1.1	Leadership and commitment to quality management system.
		5.1.2	Customer focus.
	5.2	Quality policy.	
		5.2.1	Requirements of top management with respect to quality policy.
		5.2.2	Specific requirements in respect of organization's policy.
	5.3	Organizational roles, responsibilities, and authorities.	
6.0	**Planning for Quality Management System**		
	6.1	Actions to address risks and opportunities.	
		6.1.1	Organization's context when planning for their QMS.
		6.1.2	How to address risks and opportunities.
	6.2	Quality objectives and planning to achieve them.	
		6.2.1	Enhancement and extension of ISO 9001:2008 requirements.
		6.2.2	Enhancement of ISO 9001:2008 sub-clause 5.4.2.
	6.3		Planning of changes.

(Continued)

TABLE 7.4 (Continued)
Clause Details of ISO 9001:2015

Clause No.	ISO 9001:2015			
7.0	**Support**			
	7.1	Resources.		
		7.1.1	General.	
		7.1.2	People.	
		7.1.3	Infrastructure.	
		7.1.4	Environment for the operation of processes.	
		7.1.5	Monitoring and measuring resources.	
		7.1.6	Organizational knowledge.	
	7.2	Competence.		
	7.3	Awareness.		
	7.4	Communication.		
	7.5	Documented information.		
		7.5.1	General.	
		7.5.2	Creation and updating.	
		7.5.3	Control and documented information.	
			7.5.3.1	Availability of document when needed.
			7.5.3.2	Distribution, access, and retrieve.
8.0	**Operation**			
	8.1	Operational planning and control.		
	8.2	Determination of requirements for products and services.		
		8.2.1	Customer communication.	
		8.2.2	Determination of requirements related to products and services.	
		8.2.3	Review of requirements related to products and services.	
	8.3	Design and development of products and services.		
		8.3.1	General.	
		8.3.2	Design and development planning.	
		8.3.3	Design and development inputs.	
		8.3.4	Design and development controls.	
		8.3.5	Design and development outputs.	
		8.3.6	Design and development changes.	
	8.4	Control of externally provided products and services.		
		8.4.1	General.	
		8.4.2	Type and extent of control of external provision.	
		8.4.3	Information for external providers.	
	8.5	Production and service provision.		
		8.5.1	Control of production and service provision.	
		8.5.2	Identification and traceability.	
		8.5.3	Property belonging to customers or external providers.	
		8.5.4	Preservation.	
		8.5.5	Post-delivery activities.	
		8.5.6	Control of changes.	
	8.6	Release of products and services.		
	8.7	Control of nonconforming process outputs, products, and services.		

(*Continued*)

TABLE 7.4 (Continued)
Clause Details of ISO 9001:2015

Clause No.	ISO 9001:2015		
9.0	**Performance evaluation**		
	9.1	Monitoring, measurement, analysis, and evaluation.	
		9.1.1	General.
		9.1.2	Customer satisfaction.
		9.1.3	Analysis and evaluation.
	9.2	Internal audit.	
		9.2.1	Organization's requirement to carry out internal audit.
		9.2.2	Requirements to how audit programs must be structured.
	9.3		Management review.
		9.3.1	Review of QMS by top management.
		9.3.2	Specific requirements with respect to management review.
10	**Improvement**		
	10.1	General.	
	10.2	Nonconformity and corrective action.	
		10.2.1	How the organization acts when nonconformity is identified.
		10.2.2	Documented information relating to nonconformity.
	10.3	Continual improvement.	

TABLE 7.5
List of Quality Manual Documents—Consultant

List of Quality Manual Documents—Consultant (Design and Supervision)

Document No.	Title Quality Document	Relevant Clause in 9001:2015
QC-0	Circulation list	
QC-00	Records of revision	
QC–1.1	Understanding the organization and its context	4.1
QC–1.2	Monitoring and review of internal and external issues	4.1
QC–2.1	Relevant requirements of stakeholders	4.2
QC–2.2	Monitoring and review of stakeholder's information	4.2
QC-3	Scope of quality management system	4.4
QC-4	Project quality management system	4.4
QC-5	Management responsibilities	5.1
QC-6	Customer focus	5.1.2
QC-7.1	Quality Policy (Organization)	5.2
QC-7.2	Quality Policy (Project)	5.2.2
QC-8	Organizational roles, responsibilities, and authorities (Organization chart)	5.3
QC-9	Preparation and control of project quality plan	6.0

(Continued)

TABLE 7.5 (Continued)
List of Quality Manual Documents—Consultant

List of Quality Manual Documents—Consultant (Design and Supervision)

Document No.	Title Quality Document	Relevant Clause in 9001:2015
QC-10.1	Project risk (during design)	6.1
QC-10.2	Project risk (during construction)	6.1
QC-11	Project quality objective	6.2
QC-12.1	Change Management (During design phase)	6.3
QC-12.2	Change Management (During construction phase)	6.3
QC-13	Office resources (human resources, office equipment, design software)	7.1
QC-14	Infrastructure	7.1.3
QC-15	Work environment	7.1.4
QC-16	Human resources (design team, supervision team).	7.2
QC-17.1	Training in quality system.	7.2
QC-17.2	Training in quality auditing.	7.2
QC-17.3	Training in operational/technical skills.	7.2
QC-18	Communication—internal and external	7.4
QC-19.1	Control of documents for general application	7.5.2/3
QC-19.2	Control of documents for specific projects	7.5.2/3
QC-20	Records updates	7.5.2
QC-21	Control of quality records	7.5.3
QC-22.1	Planning of engineering design and quality plan	8.1/8.3/ 8.3.1/2/3
QC-22.2	Design development (design-bid-build)	8.1/8.3/ 8.3.1/2/3
QC-22.3	Design development (design-build)	8.1/8.3/ 8.3.1/2/3
QC-23.1	Evaluation of sub-consultant and selection	8.4
QC-23.2	Communication with sub-consultant	8.4.3
QC-24	Engineering design procedure	8.5
QC-25.1	Construction supervision procedure	8.5
QC-25.2	Project management procedure	8.5
QC-25.3	Construction management procedure	8.5
QC-26.1	Control of nonconforming work (design errors)	8.7
QC-27	Project review (management and control)	9.1
QC-28	Internal quality audits	9.2
QC-29	Management review	9.3
QC-30.1	Corrective action	10.2.2
QC-30.2	Preventive action	10.3
QC-31	Control of client complaints	10.2.2

TABLE 7.6
List of Quality Manual Documents—Contractor

List of Quality Manual Documents-Contractor

Document No.	Title Quality Document	Relevant Clause in 9001:2015	Correlated Clause in ISO 9001:2008	Version/ Revision Date
QC-0	Circulation list			
QC-00	Records of revision			
QC–1.1	Understanding the organization and its context	4.1	1.1	
QC–1.2	Monitoring and review of internal and external issues	4.1	1.1	
QC–2.1	Relevant requirements of stakeholders	4.2	1.1	
QC–2.2	Monitoring and review of stakeholder's information	4.2	1.1	
QC-3	Scope of quality management system	4.3/4	4.0,4.1	
QC-4	Project quality management system	4.4	4.1	
QC-5	Management responsibilities	5.1	5.5.1	
QC-6	Customer focus	5.1.2	5.2	
QC-7.1	Quality Policy (Organization)	5.2	5.3	
QC-7.2	Quality Policy (Project)	5.2.2	5.3	
QC-8	Organizational roles, responsibilities, and authorities (Organization chart)	5.3	5.5.1/2	
QC-9	Preparation and control of project quality plan	6.0	5.4.2	
QC-10	Project risk management	6.1	5.4.2,8.5.3	
QC-11	Project quality objectives	6.2	5.4.1	
QC-12.1	Change Management (Scope)	6.3	5.4.2	
QC-12.2	Change Management (Variation orders. Site work instructions.)	6.3	5.4.2	
QC-13	Construction Resources (human resources, equipment, and machinery)	7.1	6.1	
QC-14	Infrastructure	7.1.3	6.3	
QC-15	Work Environment	7.1.4	6.4	

(*Continued*)

TABLE 7.6 (Continued)
List of Quality Manual Documents—Contractor

List of Quality Manual Documents-Contractor				
Document No.	Title Quality Document	Relevant Clause in 9001:2015	Correlated Clause in ISO 9001:2008	Version/ Revision Date
QC-16	Control of Construction, Material, Measuring, and Test Equipment	7.1.5	7.6	
QC-17	Control of Human Resources	7.2	6.2.1	
QC–18.1	Training and Development in Quality System	7.2	6.2.2	
QC–18.2	Training in Quality Auditing	7.2	6.2.2	
QC–18.3	Training in Operational/ Technical Skills	7.2	6.2.2	
QC-19	Communication—Internal and External	7.4	5.5.3	
QC-20.1	Control of Documents for General Application	7.5.2/3	4.2.3	
QC-20.2	Control of Documents for Specific Projects	7.5.2/3	4.2.3	
QC-21	Records Updates	7.5.2	4.2.3/4	
QC-22	Control of Quality Records	7.5.3	4.2.4	
QC-24	Documents Control (logs)	7.5.3.2	4.2.3/4	
QC-25	Project Planning and Control	8.1	7.1	
QC-26	Project Specific Requirements	8.2	7.2	
QC-27	Project Specific Quality Control Plan	8.2	7.2	
QC-28.1	Tender Documents	8.2	7.2	
QC-28.2	Tender Review	8.2	7.2.1/2/3	
QC-28.3	Contract Review	8.2.1	7.2.1/2/3	
QC-29	Variation Review	8.2.3	7.2.2	
QC-30	Construction Processes	8.2.2	7.2.1	
QC-31	Engineering and Shop Drawings	8.3	7.3.1	
QC-32	Design Developments for Design-Build Projects	8.3		
QC-33	Evaluation of Subcontractors and Selection	8.4	7.4.1	

(*Continued*)

TABLE 7.6 (Continued)
List of Quality Manual Documents—Contractor

List of Quality Manual Documents-Contractor

Document No.	Title Quality Document	Relevant Clause in 9001:2015	Correlated Clause in ISO 9001:2008	Version/ Revision Date
QC-34	Communication with Subcontractors, Material Suppliers, Vendors	8.4.3	7.4.1/2/3	
QC-35.1	Inspection of Subcontracted Work	8.4.2	7.4.3	
QC-35.2	Incoming Material Inspection and Testing	8.4.3	7.4.3	
QC-36	Installation Procedures	8.5	7.5.1/2	
QC-37	Product Identification and Traceability	8.5.2	7.5.3	
QC-38	Identification of inspection and test status	8.5.2	7.5.3	
QC-39	Control of Owner-Supplied Items	8.5.3	7.5.4	
QC-40	Handling and storage	8.5.4	7.5.5	
QC-41	Construction Inspection, Testing, and Commissioning	8.6	8.2.4	
QC-42.1	Control of Nonconforming Work	8.7	8.3	
QC-42.2	Control of Nonconforming Work	8.7	8.3	
QC-42.3	Project Performance Review	9.1.1	8.1	
QC-42.4	Project Quality Assessment and Measurement	9.1.2	8.1	
QC-43	Internal Quality Audits	9.2	8.2.2	
QC-44	Management Review	9.3	5.6	
QC-45	New Technology in Construction	10.1	8.5.1	
QC-46.1	Corrective Action	10.2.2	8.5.2	
QC-46.2	Preventive Action	10.3	8.5.3	
QC-47	Control of Client Complaints	10.2.2	8.3	

REFERENCES

ANSI/ISO/ASQ Q9001:2015. *Quality Management Systems—Requirements*. Quality Press, Milwaukee, WI.

Arditi, D. and Gunaydin, H. M. (1997). Total quality management in the construction process. *International Journal of Project Management*, 15(4), 235–243.

Boiral, O. (2011). Managing with ISO systems: Lessons from practice. *Long Range Planning*, 44(3), 197–220.

Boiral, O. and Amara, N. (2009). Paradoxes of ISO 9000 performance: A configurational approach. *The Quality Management Journal*, 16(3), 36–60.

Brown, R. (1994). Does America need ISO 9000? *Machine Design*, 70–74.

Cachadinha, N. M. (2009). Implementing quality management systems in small and medium construction companies: A contribution to a road map for success. *Leadership and Management in Engineering*, 9(1), 32–39.

Chini, A. R. and Valdez, H. E. (2003). ISO 9000 and the U.S. construction industry. *Journal of Management in Engineering*, 19(2), 69–77.

Curkovic, S. and Pagell, M. (1999). A critical examination of the ability of ISO 9000 certification to lead to a competitive advantage. *Journal of Quality Management*, 4(1), 51–67.

Ekanayake, L. L. and Ofori, G. (2004). Building waste assessment score: Design-based tool. *Building and Environment*, 39(7), 851–861.

Gillies, A. (2011). Improving the quality of information security management systems with ISO27000. *The TQM Journal*, 23(4), 367–376.

Harrington, H. J. (1997). The fallacy of universal best practices. *The TQM Magazine*, 9(1), 61–75.

International Organization for Standardization (2022). The ISO survey of management system standard certification.

ISO 9001:2015/Amd 1:2024. Quality management systems—requirements—AMENDMENT 1: Climate action changes.

Kale, S. and Arditi, D. (2006). Diffusion of ISO 9000 certification in the precast concrete industry. *Construction Management and Economics*, 24(5), 485–495.

Lenssen, N. and Roodman, D. M. (1995). Making better buildings. In L. R. Brown, C. F. Flavin, H. French, L. Starke, D. Denniston, H. Kane, et al. (Eds), *State of the World, a Worldwatch Institute Report on Progress Toward Sustainable Society* (pp. 95–112). W.W. Norton, New York.

Libby, J. (2007). Adding credibility. *Manufacturing Today*, 102–103.

Martinez-Costa, M. and Martinez-Lorente, A. R. (2007a). A triple analysis of ISO 9000 effects on company performance. *Journal of Productivity and Performance Management*, 56(5–6), 484–499.

Martinez-Costa, M. and Martinez-Lorente, A. R. (2007b). ISO 9000:2000: The key to quality? An exploratory study. *The Quality Management Journal*, 14(1), 7–18.

Ollila, A. (2011). The role of quality system in performance improvement – is ISO 9001 up-to-date? *Global Conference on Business and Finance Proceedings*, 6(2), 304–314.

Quazi, H. A., Hong, C. W. and Meng, C. T. (2002). Impact of ISO 9000 certification on quality management practices: A comparative study. *Total Quality Management*, 13(1), 53–67.

Renuka, S. D. and Venkateshwara, B. A. (2006). A comparative study of human resource management practices and technology adoptions of SMEs with and without ISO certification. *Singapore Management Review*, 28(1), 41–61.

Rother, M. and Shook, J. (2003). *Learning to See*. The Lean Enterprise Institute, Brookline, MA.

Seddon, J. (1998). The case against ISO 9000. *ISO 9000 News*, 12–15.

Sowards, D. (2013). Quality is key to lean's success. *Contractor Magazine*, 44, 52.

Stevenson, T. H. and Barnes, F. C. (2001). Fourteen years of ISO 9000: Impact, criticisms, costs, and benefits. *Business Horizons*, 44, 45–51.

Symonds, J. (1998). ISO 9000-case sensitive. *ISO News*, 6, 12–15.

Tan, R. R. and Lu, Y. G. (1995). On the quality of construction engineering design projects: Criteria and impacting factors. *International Journal of Quality & Reliability Management*, 16(6), 562–574.

Turk, A. M. (2009). ISO 14000 environmental management system in construction: An examination of its application in Turkey. *Total Quality Management*, 20(7), 713–733.
Von Solms, B. and von Solms, R. (2005). From information security to . . .business security. *Computers & Security*, 24(4), 271–273.
Vrassidas, L. and Chatzistelios, G. (2014). Quality management systems development based on product systems taxonomy. *The TQM Journal*, 26(2), 215–229.
West, J. E. and Cianfrani, C. A. (2014). Managing the system revision introduces focus on organizational operating conditions. *Quality Progress*, 53–56.
Zeng, S., Tan, C., Deng, Z. and Tam, V. (2003). ISO 14000 and the construction industry: Survey in China. *Journal of Management Engineering*, 19(3), 107–115.

8 Quality Auditing

Abdul Razzak Rumane

8.1 QUALITY AUDITING

A quality audit is a formal or methodical examination, review, and investigation of an existing system to determine whether agreed-upon requirements are being met. An audit is a systematic independent and documented process to verify or evaluate and report the degree of compliance to the agreed-upon quality criteria, or the specification or contract requirements of the product, services or project. There are mainly three types of audits. These are:

1. Product Audit
2. Process Audit
3. System Audit
4. Compliance Audit
5. Adequacy Audit

To achieve a competitive advantage, effective quality improvement is critical for an organization's growth. Quality audit serves the purpose of examining the effectiveness of management-directed control programs. The audit of the existing quality system that is being implemented provides factual information to develop long-term organizational strategies for quality.

A quality audit is defined as a systematic, planned, independent, and documented process to verify or evaluate, and report the degree of compliance to the agreed-upon quality criteria, or the specification or contract requirements of the product, services, or project.

A quality audit determines whether quality activities and their outcomes comply with the plan and whether the plan is achieving its goals. Quality audits in the construction industry are used in two broad categories. The first is auditing the quality of workmanship such as code compliance. The second category pertains to the assessment of the quality management system. This is necessary to validate the system and its processes and discover any gaps that may exist.

Many quality standards stipulate that a quality system needs to be audited by third parties independent of the organization providing the system. The exception to this is when companies use internal auditors sometimes selected from within the organization itself to find the inadequacies prior to external auditors' assessment.

An auditor is a person or a team of people who have been trained in quality auditing and/or have experience in auditing. Auditors can be from the same company (internal audit), from other independent organizations (i.e., American Society for

DOI: 10.1201/9781003636687-8

TABLE 8.1
Differences between Audit and Inspection

Audit	Inspection
• Audit is a function of QA process.	• Inspection is a function of QC process.
• Audit is done while the product is being built or during the process.	• Inspection is done after completion of a particular activity, or product, or process.
• Audits often go deeper than the inspection. Audits are typically concerned with a wider range of requirements.	• Inspections generally focus on smaller number of characteristics.
• Audits are external to the process so the process can continue whether the auditor is present or finds discrepancies during the audit (exceptions include safety issues or regulatory audits, which can result in process being stopped).	• Inspections are typically a mandatory step in the process and included as such in process instructions and other workflow documentation (routing lists, and so on).
• Audis are less frequent than Inspection.	• Inspections are typically more frequent than audits.
Similarities	
Audits and inspections can both be performed on products, processes, projects, or systems.	
Audits and inspections both typically assess compliance with some standard.	

Quality (ASQ)), government agencies (Nuclear Regulatory Agency (NRA)) or even auditors sent by clients who are interested in learning more about the organization's quality system and whether or not it is effective.

8.1.1 Auditing versus Inspection

TABLE 8.1 lists major differences between audit and inspection.

8.2 CATEGORIES OF AUDITING

Assessment of quality activities can be evaluated from two perspectives:

1. Assessment that focuses on customer satisfaction results but includes an evaluation of the present quality system.
2. Assessment that focuses on the evaluation of the present quality system with little emphasis on customer satisfaction results.

In either case, the assessment can be performed by the organization itself or by an external body. The assessment performed by the organization is known as "self-assessment," whereas an assessment performed by an external body is referred to as a quality audit. A quality audit is a formal or methodical examination, review, and investigation of existing systems to determine whether agreed-upon requirements are being met.

Audits are mainly classified as:

First Party—Audit your own organization (Internal Audit).
Second Party—Customer audits the supplier.
Third Party—Audits performed by independent audit organization.

8.2.1 First Party Auditing

First-Party audits are usually conducted by employees of the company who are experts on assessment. They usually are selected from other parts of the company to minimize conflict of interest. Experts from outside may be called in to conduct audits. Most frequently are volunteer members of the local quality association who come for anywhere from one to three days to conduct audits. Those auditors have some type of quality certification to assess the system of the organization. At the conclusion of the audit, the lead auditor will present the audit findings to the leadership of the organization including staff in charge of quality compliance.

8.2.2 Second Party Auditing

Second-Party audits are usually external audits performed by a party outside the organization. These audits are usually conducted when a prospective client is interested in assessing the quality system of the supplier prior to a purchase agreement. These audits could range from being process-specific to an overall system audit. Clients or customers could waive such audits (or limit them) if the organization is already certified to an ISO 9001 or similar management system standard.

8.2.3 Third Party Auditing

Third-Party auditors are conducted by agencies or certification and/or accreditation bodies. This could be government or regulatory agencies such as the Food and Drug Administration (FDA) in the United States. It could also be auditing services through independent entities specialized in quality system certification. A key feature of a certification system is that inspections are carried out by independent bodies (third-party certification) in accordance with standards laid down by external organizations.

There are many certification bodies worldwide that provide certification services for the ISO 9001 or other management system standards. If the audit is successful, then the certification body (CB) such as BSI, TUV, and NQA, among others, awards a certificate to be renewed every three years through a recertification audit. Between the three-year cycle, surveillance audits are conducted annually to ensure the QMS is operating effectively. The surveillance audits are not as comprehensive as the certification or re-certification audits. If the organization being audited has multiple locations, then surveillance may be conducted on some every year so that all are audited within a certification cycle.

Sometimes firms opt for a pre-audit before the audit. Due to the high cost of third-party certification, often a pre-audit is done to ensure that the company is well prepared and equipped before the major third-party certification audit.

8.3 AUDITING STANDARDS

The internal audit Clause (Section) 9.2 in the ISO 9001:2015 standard starts by requiring any organization seeking compliance with the standard to conduct internal audits "at planned intervals." In other words, the organization must plan its audits and do so on a regular basis. However, the exact frequency of performing these audits is not specified and is left for the individual organization to decide. Based on the author's experience, most companies tend to audit their entire Quality Management System (QMS) every year with fewer opting to do so on either a higher or lower frequency. This doesn't mean that the frequency could not change over time. An organization may start with an annual frequency and switch to a semiannual basis after a few years. They may also audit certain areas more frequently than others. For example, it is not uncommon to audit areas such as "control of nonconforming output" or "corrective action" at a higher frequency than "purchasing," for example.

As indicated above, the ISO 9001:9015 standard further explains that the objective of such audits is to provide information on whether the QMS conforms to requirements. These requirements could originate from:

- The ISO 9001 standard itself.
- Other requirements not mentioned in the standard but are essential of operations.
- Applicable regulatory and statutory requirements.

Conformance to requirements is an indication that the quality management system is comprised of documented processes to address them adequately. For example, handling incidents of nonconformance generated internally may be addressed in the form of general procedures or a combination of documented information. Such a procedure or other documented information must cover all aspects of handling a nonconforming product (or service) including identification, containment, and disposition. Furthermore, a procedure that includes provisions to handle requirements must be communicated to all affected personnel. The communication may take the form of official training on how to handle nonconforming output.

Having provisions such as procedures in place does not mean that things get done or get done correctly. How do we know that affected employees are following established processes? In the nonconforming output example, how can we know that the responsible associate is identifying a nonconforming product or service? As indicated in the last part of Clause (Section) 9.2.1 of the ISO 9001 standard, the internal audits are also there to ensure that the provisions established through procedures, instructions, etc., are implemented effectively and maintained. This can be verified through records, observing work being done, and interviews with employees. For nonconforming output, records include reports of disposition and status tags attached to nonconforming products or outputs. In a manufacturing environment, an area for nonconforming or suspect products is often created to place the suspect material until disposition. In construction projects, not using the correct materials by a contractor to complete a task on the project schedule can be viewed as a case of non-conformance. Dealing with such a nonconformance may take the form of informing the client and making corrections. It may also require corrective action if it is a repeated problem.

8.3.1 ISO Standards for Auditing

While some form of internal auditing is included as a requirement in each contractual management system standard, no guidelines within these standards are given on how to set up and implement audit programs. That is, standards like ISO 9001, ISO 14001, and ISO 13485 require that the organization perform internal audits on a regular basis with slight differences. However, the sections stipulating such requirements do not elaborate on how to set up such programs. Thus, guidelines on auditing management systems were needed.

The International Organization for Standardization first published guidelines on auditing management systems in 2002. These guidelines were only applicable to quality and environmental management systems. As the number of management system standards grew over the years including, but not limited to, ISO 13485 (medical device), and ISO 22000 (food safety), broader guidelines for auditing were needed. Most of these standards have a common structure and similar base requirements. As a result, ISO 19011 guidelines were updated in 2011 and then in 2018 to be applicable to all management system standards.

8.3.1.1 ISO 19011:2018—Guidelines for Auditing Management Systems

The latest revision of ISO 19011 published in the Summer of 2018 provides guidelines to plan and conduct audits for any management system. The title of ISO 19011 is Guidelines for Auditing Management Systems, it is not a contractual standard. This means that these are just guidelines and not requirements for any type of certification. However, most auditor training programs, from internal to third party, use this standard as the basis of their body of knowledge.

The ISO 19011 standard can be used by all sizes and types of organizations and for any type of audit. However, it does concentrate on internal auditing and second-party audits. That is, organizations can use this standard to set up their audit programs for their organizations.

ISO 19011 is based on several clauses starting with the following three informational ones: Scope of the standard including applicability issues, normative references, and definitions of key terms used in the body of the standard. The rest of the clauses (4 through 7) include the actual guidelines for auditing management systems:

- Principles of Auditing
- Managing Audit Programs
- Conducting Audits
- Auditor Competence and Evaluation

The principles of auditing are considered the pillars for making audits effective and reliable. When adhered to, these principles ensure that similar conclusions can be reached by a number of auditors working independently of each other and auditing the same process or system. At a glance, these principles seem logical and should be easy to adhere to. However, in practice, audit programs may turn into a paperwork exercise if auditees do not cooperate for one reason or another. Therefore, when auditors receive their training, examples, and scenarios of how such principles could be compromised should be worked on.

The first of the auditing principles is integrity. Oxford Dictionary defines integrity as "The quality of being honest and having strong moral principles" and "The condition of being unified or sound in construction." For auditors, this means that they should do their work ethically following the code of ethics presented by the granting entity. For the American Society of Quality, which grants certification for quality auditors, certified auditors must adhere to a code of ethics, besides being the right thing to do in the first place.

Auditors should also report on findings truthfully and accurately without omitting significant findings or exaggerating others. This should be done with due diligence and the ability to make decisions in all types of auditing situations.

Confidentiality is another principle that many organizations are increasingly sensitive to. Oxford Dictionary defines confidentiality as "The state of keeping or being kept secret or private." It is common for many organizations to require signing a confidentiality agreement before accessibility to information is granted. Auditors are exposed to all types of company information which must be protected. This includes propriety, security-related, or personnel information, among others. If auditors need information to examine later at an offsite location, they usually require permission to do so, in addition to the confidentiality agreement.

One of the fundamental principles of auditing is independence. That is, the auditor must be independent of the function or process being audited. The reason for this is obvious; when the rule of independence is violated, impartiality and objectivity of results may be compromised. For example, someone from procurement or the purchasing department should not audit the process of qualification and evaluation of external providers. The question is how independent should an auditor be? This question comes up often, particularly when there are many functions within a department. In the procurement example, what if the auditor does not directly work with external providers (suppliers or contractors)? Does that mean he/she is independent? The answer to this question depends on the level of involvement and whom you ask. However, a good rule of thumb is not to audit functions within the same department or under the same manager. It is much harder to do so in smaller organizations where one employee could be responsible for multiple functions. In such cases, management should make every effort to ensure objectivity, including subcontracting audits by an external source if necessary.

Another principle of importance is being fact-based with objective evidence. Audit findings should not be based on opinions and the facts stated must be verifiable. This means that auditors should reference the samples taken with enough information that, if needed, someone can return to it later and find such evidence. Additionally, auditors should take enough samples within the period of interest, including current ones, to ensure reaching reliable conclusions.

Since contractual management standards, such as ISO 9001, now include risk-based thinking, ISO 19011 also includes reference to risks and opportunities under its principles. It states that "the risk-based approach should substantively influence the planning, conducting, and reporting of audits in order to ensure that audits are focused on matters that are significant to the client."

ISO 19011 also addresses the management of audit programs. According to the standard, an audit program is defined as "arrangements for a set of one or more audits planned for a specific time frame and directed towards a specific purpose." In practice, audit programs are established and maintained for ongoing management system audits. One audit program may be established for one or more management systems such as environmental and quality. In any case, the extent of an audit program depends on the size and complexity of the organization. For example, an organization with multiple sites and outsourcing vendors or contractors requires more details and resources than a simple manufacturing process. Auditing a multi-site construction project with vendors, consultants, and contractors is an example of higher complexity. For this, the audit program includes auditing both internal processes as well as external providers (e.g., contractors). Priorities should be given to processes and entities that pose higher risks for achieving organizational and project objectives.

Managing audit programs involves establishing audit objectives, evaluating risks, identifying responsibilities for those involved, selecting auditors, managing program records, and evaluating and improving the auditing process as well as the overall audit program.

The first step in establishing an audit program is establishing its aims. This will help in determining the required resources, including auditors. The following questions can be asked to help determine the overall objectives:

- What are the needs and expectations of clients/customers?
- What are the needs and expectations of other interested parties including employees (e.g., internal auditors) and external providers (e.g., contractors), among others?
- What are the competencies needed for those managing the audit program?
- What are the current issues and risks with the processes being audited?
- What are the results of previous audits conducted on each of the processes?

The objectives for an audit program will vary from organization to organization. For a construction/engineering organization, the objectives could include the following:

- Identifying current shortcomings of different processes (nonconformance) regarding meeting quality management system requirements as well as contract requirements.
- Identifying opportunities for improvement.
- Identifying capabilities of established processes and external providers. In other words, are the processes capable of delivering to customer requirements consistently? In construction projects cost and time represented by budget and schedule, respectively, are key attributes for successful project completion.

- Complying with legal and statutory requirements.
- Aligning project objectives with overall organizational objectives.

The program should incorporate and monitor these objectives over time to assess the role of auditing in achieving them. After all, the audit program is a part of a process-performance evaluation so countermeasures can be implemented and improvements can be made.

Another important aspect of establishing an audit program is the selection of auditors. In addition to including an appropriate number of auditors to cover the scope, it is essential to consider factors such as competence and knowledge of the processes being audited and availability. In most cases, internal QMS auditors have other functions within the organization that limit their availability. Some organizations supplement their internal audit team with subcontracted external auditors.

Generally, organizations certified to ISO 9001 and/or other management system standards have a procedure for their QMS audit program. Such a procedure covers all aspects of conducting audits, including:

- Auditor Selection
- Audit Planning
- Audit Scheduling
- Conducting Audits
- Audit Reporting
- Audit Follow-Up

TABLE 8.3 illustrates the typical procedure for internal QMS audits.

Before conducting audits, it is imperative for the audit program manager to consider planning factors such as auditee availability, audit feasibility, and resources,

TABLE 8.2
Typical Procedure for QMS Audits

Purpose and Scope

This procedure describes the process of planning, conducting, and reporting internal quality audits.

Responsibilities

The program manager is responsible for ensuring that internal audits are conducted according to schedule and for reporting to management on the status of the quality management system as a result of internal audits. Qualified auditors will conduct audits and report on findings.

Procedure

1	Auditor Selection: Auditor must be selected by the program manager according to the following criteria: a. Training: Each auditor must have had responsible training in auditing. This training can be provided internally or externally. b. Independence: An auditor cannot audit his/her immediate areas of work. Auditors can also be subcontracted from an outside source. c. Knowledge: auditors must be scheduled based on their knowledge and experience in the function being audited.

(Continued)

TABLE 8.2 (Continued)
Typical Procedure for QMS Audits

2	Audit Planning: Audits must be planned so that the entire quality system is audited each year. Items that indicated problems in prior audits or other nonconformities should be included in current audits. The program manager should inform auditors of items to be included as part of their checklists or provide them with applicable checklists.
3	Audit Scheduling: The program manager prepares the audit schedule quarterly. Each audit schedule must at least include: a. Area to be audited. b. Audit dates. c. Auditor's name. d. Special instructions, if any.
4	Audit Findings: Auditors perform audits through questioning, observing, and verifying records. The following findings must be recorded: a. Area for improvement: This occurs when the system shows some weakness but not necessarily nonconformance to establish requirements. These weaknesses may lead to nonconformance if not treated. b. Minor nonconformance: This occurs when there is only a lapse in the system given that that requirement shows conformance most of the time. c. Major nonconformance: This occurs when there is a breakdown in the system. It is also possible that few minors within the same area at one time could constitute a major.
5	Audit Reporting: a. The auditee must acknowledge the findings by signing the specified space on the internal audit report. If there is a conflict, the program manager should resolve it. b. A copy of the audit sheet should be given to the auditee after getting the signature. c. The program manager issues formal corrective action requests to the auditee for any nonconformance immediately by using the nonconformance database. d. The areas of improvement should be left for the auditee to decide on when to start such a process of improvement. Records for such improvements must be kept with the auditee for future reference. e. The program manager provides management with reports on status of the quality management system regarding the number of nonconformances, frequencies, trends, etc., and corrective action taken. f. Nonconformance and areas for improvement uncovered may be included as items in subsequent audits.
Records	Audit Schedules Audit Reports

among others. TABLE 8.3 can be used as a checklist for this purpose (audit preparation). Once a checklist is completed, an action list could be prepared for items not addressed or completed.

The last of the guidelines addressed by the ISO 19011 standard is the competence and evaluation of auditors. These guidelines provide details on factors that should

TABLE 8.3
Checklist for Audit Preparation

Serial Number	Preparation Item	Y/N/NA
1	Auditee contacted with details?	
2	Auditee confirmed audit times and dates?	
3	Guides can be available if needed.	
4	Auditee confirmed opening and closing meeting times and dates?	
5	Auditee communicated back confidentiality requirements?	
6	Auditee communicated back safety and health requirements?	
7	Auditor independent of the process being audited?	
8	Auditor is knowledgeable of process being audited?	
9	Criteria for audits (procedures, contracts, and standards) are available for review.	
10	Adequate time is allowed?	
11	Other resources (e.g., computer access) are available to perform the audit?	
12	Risk audit poses on audit process, if any, are considered?	
13	Sampling process of objective evidence is determined?	
14	Working language of the auditee is manageable by the auditor?	

be considered in selecting an audit team. To do so, the standard suggests a variety of methods for evaluating potential auditors which are summarized in Table 8.3. Generally, the program manager may use the following steps in selecting auditors:

1. Identify competence needs: This includes both auditing skills and knowledge about processes being audited. So, competence would include personal behavior attributes as well as skills and knowledge. Many organizations select potential internal auditors based on interest and personal behavior, among other factors. For skills and knowledge, appropriate training is provided.
2. Prepare evaluation criteria to make the selection. The criteria could be selected from different departments (distribution), years of experience, skills and knowledge, and personal behavior, among others.
3. Use the proper evaluation methods: This should include at least two from Table 8.3 as recommended by the standard.
4. Conduct the evaluation, keeping in mind that some may need additional training before being selected.
5. Maintain the current auditors' list and provide continual opportunities for skill improvement and professional development.

One of the new aspects of auditing in the ISO 19011:2018 standard is the inclusion of provisions for auditing from a remote location. In addition to traditional onsite methods, remote audit methods through means such as video conferencing include:

- Interviews.
- Observing work being completed.
- Conducting document reviews with and without auditee involvement.
- Analyzing data presented remotely by the auditee.

8.4 QUALITY AUDITING PROCESS IN CONSTRUCTION PROJECTS

Quality audit is a systematic, planned, independent, and documented process to verify or evaluate and report the degree of compliance to the agreed-upon quality criteria, or the specification or contract requirements of the product, services, or project. FIGURE 8.1 illustrates the PDCA cycle for the auditing process. FIGURE 8.2 illustrates the typical quality auditing process for construction projects and TABLE 8.4 lists typical contents of the audit plan.

8.4.1 Auditing Performance Criteria

Quality auditing in construction projects is a systematic and planned examination, evaluation, assessment, review of project management processes, project activities, project deliverables, systems, products, and project status/compliance with respect to the required project performance by an internal or external auditor for providing the

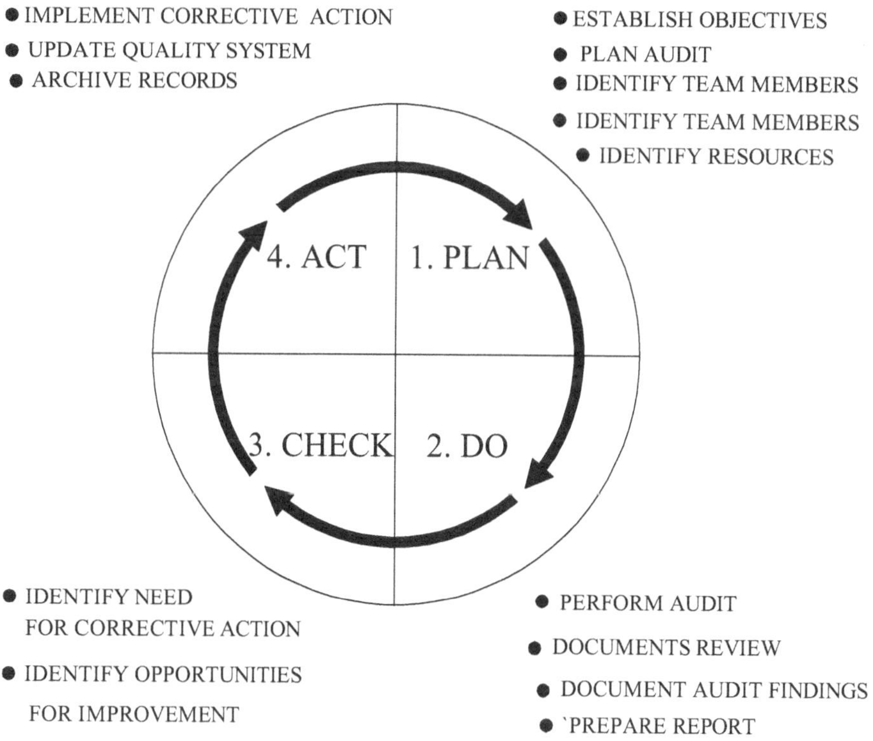

FIGURE 8.1 PDCA cycle for the auditing process.

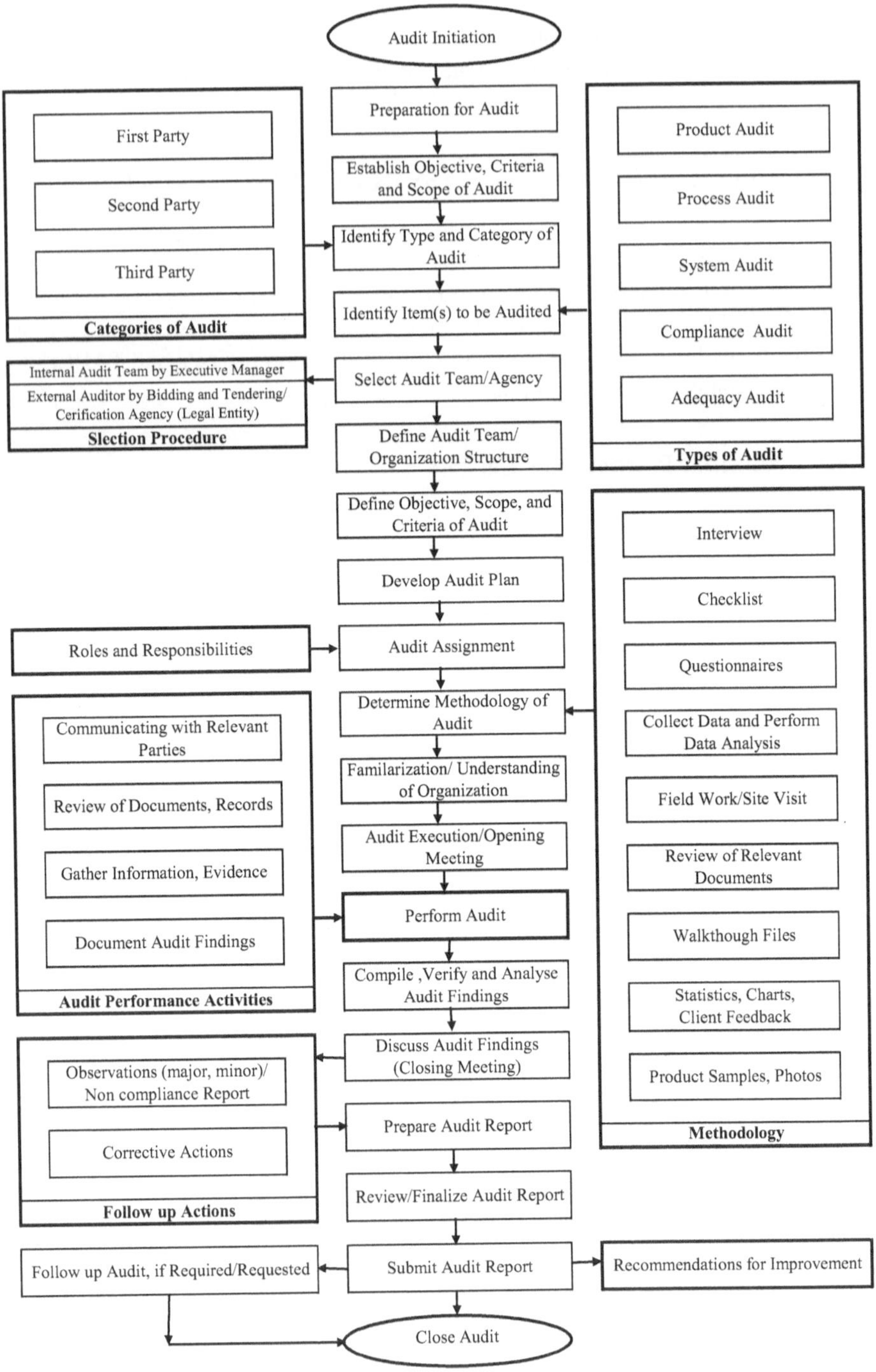

FIGURE 8.2 Typical quality auditing process for construction projects.

TABLE 8.4
Contents of Audit Plan

Section	Topic
1	Purpose and Objective
2	Scope of Audit
3	Applicable Documents
4	Audit Schedule
	4.1 Date, Duration
	4.2 Location
5	Audit Methodology, Tools
6	Logic Arrangements
7	Team Members
	7.1 Auditor
	7.2 Auditee
8	Roles and Responsibilities, Assignment
9	Allocation of Resources
10	Confidentiality Requirements
11	Distribution of Report
12	Follow up Action

information to those (clients) in need of assurance in respect to their defined scope to achieve goals and objectives of the project.

In order to plan, manage, and control construction projects, the project life cycle is divided into a number of phases depending on the construction processes, technologies, processes, and complexity. The audit process can be structured and performed in accordance with individual phases.

Audit performance is the fieldwork carried out by the auditor at the audit location. It mainly consists of the following activities:

- Opening meeting.
- Communication with relevant parties that will be involved in the audit.
- Understanding the system and process to be followed.
- Understanding of audit purpose and objectives.
- Confirmation of audit plan.
- Understanding of audit methodology.
- Review of documents and records.
- Gathering of information; data.
- Gathering of evidence.
- Communication with team members.
- Interview with auditee.
- Document audit findings.
- Compiling, verifying, and analysis of audit findings.
- Closing meeting.
- Preparation and submission of audit report.

Depending on the objectives of audits the construction project audit can be performed as follows:

- During the execution of each individual phase.
- After completion of each phase.
- After completion of the whole project.

The audit for construction project life cycle phases can be performed by first party, third party, and for certain activities by the second party.

8.5 QUALITY AUDITING PROCESS IN THE PHASES OF CONSTRUCTION PROJECTS

Most construction projects are custom-oriented, having a specific need and a customized design. It is always the owner's desire that his project should be unique and better. There are innumerable processes that make up the construction process, A systems engineering approach to construction projects helps to understand the entire process of project management and to manage and control its activities at different levels of various phases to ensure the timely completion of the project with economical use of resources to make the construction project most qualitative, competitive, and economical. In order to process the construction project in an effective and efficient manner and to improve control and planning, construction projects are divided into various phases. The life cycle of a major construction project has seven phases. Each phase is further subdivided into the work breakdown structure principle to reach a level of complexity where each element/activity can be treated as a single unit that can be conveniently managed. Construction projects, being unique and nonrepetitive in nature, need specified attention to maintain quality. Each project has to be designed and built to serve a specific need.

Construction project life cycle phases have construction-related technical activities that have to be assessed, verified, reviewed, and evaluated to achieve a successful project. The quality audit process for each of these phases depends on the objectives of the audits. The audit process of each phase should cover all the areas/activities to achieve audit objectives.

The quality audit objectives of construction projects vary as per clients' requirements and to what extent the details are to be assessed and verified. Some of the objectives of the audit of design phases are as follows:

- Project goals and objectives have been properly defined.
- Terms of Reference (TOR) are properly established.
- Design management activities are adequately performed and managed.
- Project management processes relating to design are properly followed.
- Quality management procedures are fully complied.
- Design deliverables are identified and produced.
- Construction project quality elements and constraints are considered while developing the design and construction documents.
- Risk factors are considered.

The following sections discuss the auditing processes for the design phases.

8.5.1 Auditing Process for Conceptual Design Phase

The conceptual design phase commences once the need is recognized. The conceptual design phase basically has two stages:

8.5.1.1 Study Stage

The study stage activities include:

- Identification of need by the owner, and establishment of main goals.
- Feasibility study which is based on owner's objectives.
- Identification of project delivery system.
- Identification of project team by selecting other members and allocation of responsibilities.
- Identification of alternatives.
- Selection of preferred alternative.
- Time schedule.
- Financial implications based on estimation of life cycle cost of the favorable alternative.
- Resources.
- Project risk.
- Authorities' approval.
- Terms of Reference (TOR).

8.5.1.2 Concept Design

- Development of concept design.

The most significant impacts on the quality of the project begin during the conceptual phase. This is the time when specifications, statement of work, contractual agreements, and initial design are developed. Initial planning has the greatest impact on a project because it requires the commitment of processes, resources schedules, and budgets. A small error that is allowed to stay in the plan is magnified several times through subsequent documents that are second or third in the hierarchy.

8.5.1.3 Auditing Process for Study Stage

The main objectives to performing a quality audit for the study stage of the conceptual design phase are:

1. Evaluate and asses process compliance with the organizational strategic policies and procedure.
2. Assess conformance to QMS and processes to meet established quality system requirements.
3. Review the process used to prepare the business case and justification for the project.

4. Examine that key risk factors are analyzed and managed to establish the business case.
5. Validate the expected schedule and cost of the project.
6. Asses that all the major activities are adequately performed and managed.
7. Assess the proper functioning of the project management system.
8. Identify if there is any weakness in the current state of the organization to manage and execution of the project.
9. Assess that the project will deliver the expected benefits.

TABLE 8.5 illustrates the audit methodology for the assessment of the study stage of the conceptual design phase.

TABLE 8.5
Audit Methodology for Assessment of Study Sage

Serial Number	Item to be Assessed	Yes	No	Comments
A. Checklist				
1	Whether the project need is established taking into consideration SMART (specific, measurable, attainable, realistic, time bound) concept?			
2	Is the project as per organization's strategic policy/ strategic plans?			
3	Is the project mandated by the management in support of a specific objective?			
4	Has the business need been clearly defined?			
5	Whether risk factors are considered prior to establish project need.			
6	Is the project conforming with government regulations?			
7	Whether assessment/analysis of need carried out as per the approved guidelines?			
8	Whether feasibility study is performed in following areas:			
	a. Technical			
	b. Economical			
	c. Financial			
	d. Market demand			
	e. Time Scale			
	f. Resource Availability			
	g. Risk Analysis			
	h. Environmental			
	i. Ecological			
	j. Sustainability			
	k. Political			
	l. Legal			
	m. Social			

(*Continued*)

TABLE 8.5 (Continued)
Audit Methodology for Assessment of Study Sage

Serial Number	Item to be Assessed	Yes	No	Comments
9	Has the business case clearly defined after considering the following points as a minimum: a. Project Need b. Stakeholders c. Project Benefits d. Financial Benefits e. Estimated Cost f. Estimated Time g. Justification			
10	Whether life cycle costing of the project has been carried out and considered?			
11	Whether project goals and objectives are established on SMART concept considering the following information: a. Project Scope and Project Deliverables b. Preliminary Project Schedule c. Preliminary Project Budget d. Specific Quality Criteria the Deliverables Must Meet e. Type of Contract to be Employed f. Design Requirements g. Regulatory Requirements h. Potential Project Risks i. Environmental Considerations Logistic requirements.			
12	Whether preferred alternatives selected after analyzing and evaluating each of the identified alternatives?			
13	Whether preferred alternatives approved by the owner/end user?			
14	Whether high-level requirements (scope, schedule, cost, and quality) for the project established?			
15	Whether environmental authority's clearance certificate obtained?			
16	Whether project delivery system is selected as per strategic decision and stability of appropriate system?			
17	Whether selected contracting/pricing is beneficial to the owner?			
18	Whether authorities' approval obtained?			
19	Whether problem/challenge identified and resolved?			
20	Whether project team is selected to suit the adopted project delivery system?			

(Continued)

TABLE 8.5 (Continued)
Audit Methodology for Assessment of Study Sage

Serial Number	Item to be Assessed	Yes	No	Comments
21	Whether project team selected as per organization's bidding and tendering process?			
22	Whether the terms of reference (TOR) cover all the requirements to fulfill the owner's project objectives and goals?			
23	Whether project boundaries and basic parameters clearly defined?			
24	Whether the project is approved by all the concerned stakeholders?			

B. Interview Questionnaires

Serial Number	Description	Response	Comments
1	Why was the project initiated?		
2	Was the project created to meet market demand?		
3	Whether the project was mandated by regulatory authority or as part of vision statement of the organization?		
4	Was there any need for such project?		
5	Whether project duration is as per organization's expected requirements?		
6	Whether project cost was within the reach of the organization?		
7	Whether there was delay in getting authorities approval?		
8	Was there any difficulty in approving the project by the concerned authority/team/owner?		

8.5.1.4 Auditing Process for Concept Design

The main objectives of performing a quality audit of a concept design of the conceptual design phase are:

1. Evaluate and assess process compliance with the organizational strategic policies and procedures.
2. Assess conformance to QMS and processes meet established quality system requirements.
3. Review the process used to prepare the concept design.
4. Verify that the concept design complies with project goals and objectives and fully meets the owner's requirements.
5. Check compliance of TOR requirements.
6. Review the process used to prepare the design report and model.
7. Validate the estimated schedule and cost of the project meeting the budget.
8. Validate compliance of quality standards and codes.

9. Check for compliance with regulatory requirements.
10. Check for sustainability considerations in the development of design.
11. Check that conservation of energy requirements are considered.
12. Check that facility management requirements are considered.
13. Review that key risk factors are analyzed and managed to develop concept design.
14. Asses that all the major activities are adequately performed and managed.
15. Assess proper functioning of the project management system.

TABLE 8.6 illustrates the audit methodology for the assessment of concept design of the conceptual design phase.

TABLE 8.6
Audit Methodology for Assessment of Concept Design

Serial Number	Item to be Assessed	Yes	No	Comments
C. Checklist				
1	Whether scope of work for development of concept design established based on TOR requirements?			
2	Whether organization chart with responsibilities to develop concept design established?			
3	Whether the concept design supports owner's project goals and objectives?			
4	Whether the owner's requirements are fully considered?			
5	Whether the concept design meet all the elements specified in TOR?			
6	Whether the design comply with organizational strategic policies?			
7	Whether the design is based on specified quality standards and codes?			
8	Does the design meet all the performance requirements?			
9	Whether the data and other information is gathered prior to start of design?			
10	Whether owner's preferred requirements collected to prepare the design and take care of the requirements in the design?			
11	Whether regulatory/statutory requirements considered?			
12	Whether the design team have specific qualifications and experience to design such project?			
13	Whether project schedule is practically achievable?			

(Continued)

TABLE 8.6 (Continued)
Audit Methodology for Assessment of Concept Design

Serial Number	Item to be Assessed	Yes	No	Comments
14	Whether project cost properly estimated?			
15	Whether concept design prepared taking into consideration project quality requirements?			
16	Whether design is coordinated for requirements of all trades?			
17	Whether designer has considered availability of resources during entire project life cycle?			
18	Whether design risks have been identified, analyzed, and responses planned for mitigation?			
19	Whether health and safety requirements considered in the design?			
20	Whether environmental constraints considered?			
21	Whether the design meet LEED requirements?			
22	Whether cost-effectiveness over the entire project life cycle is considered?			
23	Whether the reports is prepared to meet TOR requirements?			
24	Is energy conservation considered?			
25	Is sustainability considered in design?			
26	Whether all reasonable alternative options/ systems are considered for design?			
27	Whether all reasonable alternative options are considered for project economy?			
28	Whether constructability has been taken care?			
29	Whether designer used quality function deployment (house of quality) technique to develop concept design?			
30	Whether the model meets the design objectives?			
31	Does the design support proceeding to next design stage?			
32	Whether drawings were reviewed and checked for quality compliance prior to submission?			
33	Whether reports and documents reviewed prior to submission?			
34	Whether the reports are complete and include adequate information about the project?			
35	Whether the report is properly formatted and has a table of contents for each report?			
36	Does the designer have considered and followed quality management process activities related to concept design?			
37	Whether the design drawings, reports, and other documents properly labeled and archived?			

(Continued)

TABLE 8.6 (Continued)
Audit Methodology for Assessment of Concept Design

Serial Number	Item to be Assessed	Yes	No	Comments
D. Interview Questionnaires				
Serial Number	**Description**	**Response**		**Comments**
1	Whether there was any delay to collect the data?			
2	Whether there was any delay in collection of owner's preferred requirements?			
3	What method was adopted to collect owner's requirements and compliance with all the functional, parking, landscape areas?			
4	Whether concept design was prepared within the agreed-upon schedule?			
5	Whether work progress schedule was prepared to monitor the progress?			
6	Whether a timesheet was maintained to record the time spent by each staff engaged for preparation of concept design?			
7	Was there any delay in getting authorities approval?			
8	Whether risk factors are considered while developing the concept design?			
9	Whether concept design submitted in time for owner's review?			
10	Whether numbers of sets were prepared as per TOR and submitted?			
11	Whether the owner issued concept design acknowledgement letter?			
12	Does the concept design have any major comments from the owner?			
13	Whether the concept design was not approved by the owner?			
14	Whether the reasons (causes) for comments, non-approval, if any, analyzed and corrective/ preventive action taken?			

8.5.2 Auditing Process for Schematic Design Phase

Schematic/preliminary design is the basic responsibility of the architect (designer/ consultant or A&E). The central activity of schematic/preliminary design is the architect's design concept of the owner's objective which can help make detailed engineering and design for the required project/facility. Schematic/preliminary design is a subjective process of transforming ideas and information into plans, drawings, and specifications of the facility to be built. Component/equipment configurations, material specifications, and functional performance are decided during this phase.

At the preliminary design stage, the scope must define deliverables, that is, what will be furnished. It should include a schedule of dates for delivering drawings, specifications, calculations, and other information, forecasts, estimates, contracts, materials, and construction.

The main objectives of performing a quality audit of the schematic design phase are to:

1. Evaluate and assess process compliance with the organizational strategic policies and procedures.
2. Assess conformance to QMS and processes to meet established quality system requirements.
3. Review the process used to prepare the schematic design.
4. Check that the schematic design complies with project goals and objectives.
5. Verify that design deliverables are as per TOR requirements.
6. Validate the expected schedule and cost of the project meeting the project boundaries and defined parameters.
7. Validate compliance of quality standards and codes.
8. Assess compliance with the design quality management system.
9. Check for compliance with regulatory requirements.
10. Check for sustainability considerations in the development of design.
11. Check that conservation of energy requirements are considered.
12. Check that facility management requirements are considered.
13. Review that key risk factors are analyzed and managed to develop schematic design.
14. Review preliminary specifications and contract documents.
15. Asses that all the major activities are adequately performed and managed.
16. Assess the proper functioning of the project management system.
17. Check the value engineering study is performed.

TABLE 8.7 illustrates the audit methodology for the assessment of schematic design.

8.5.3 Auditing Process for Design Development/Detailed Design Phase

Design development/detailed design is the enhancement of the work carried out during the schematic design phase. During this phase, a comprehensive design of work with a detailed work breakdown structure of the design, drawings, specifications, and contract documents are prepared. The design development phase is the realm of design professionals, including architects, interior designers, landscape architects, and several other disciplines such as civil, mechanical, electrical, and other engineering professionals as needed.

During this phase detailed plans, sections, and elevations are drawn to scale, principle dimensions are noted and design calculations are checked to confirm the accuracy of the design and its compliance to the codes and standards.

TABLE 8.7
Audit Methodology for Assessment of Schematic Design

Serial Number	Item to be Assessed	Yes	No	Comments
E) Checklist				
1	Whether scope of work for development of schematic design established based on TOR requirements?			
2	Whether organization chart with responsibilities to develop schematic design established?			
3	Whether the schematic design supports owner's project goals and objectives?			
4	Whether the design include all the functional, external works, parking, and landscape areas?			
5	Whether the design comply with organizational strategic policies?			
6	Whether the owner's requirements are fully considered?			
7	Whether comments on concept design taken care while preparing schematic design?			
8	Whether the schematic design meet all the elements specified in TOR?			
9	Whether regulatory/statutory requirements considered?			
10	Whether the design is based on specified quality standards and codes?			
11	Does the design meet all the performance requirements?			
12	Does the designer consider project boundaries?			
13	Whether constructability has been taken care?			
14	Whether technical and functional capability considered?			
15	Was sustainability considered in design?			
16	Whether the design meet LEED requirements?			
17	Was energy conservation considered?			
18	Whether designer has considered facility management requirements?			
19	Whether project objectives with respect of time, cost, and quality considered while developing the design?			
20	Whether valid assumptions and constraints considered?			
21	Whether project schedule is practically achievable?			
22	Whether project cost is properly estimated?			

(*Continued*)

TABLE 8.7 (Continued)
Audit Methodology for Assessment of Schematic Design

Serial Number	Item to be Assessed	Yes	No	Comments
23	Whether schematic design prepared taking into consideration project quality requirements?			
24	Whether designer has considered availability of resources during entire project life cycle?			
25	Whether the design team have specific qualifications to design such project?			
26	Whether design risks been identified, analyzed, and responses planned for mitigation?			
27	Whether design was fully coordinated with all the trades?			
28	Whether health and safety requirements in the design are considered?			
29	Whether design conform with fire and egress requirements?			
30	Whether environmental constraints and requirements considered?			
31	Whether specification prepared taking into consideration international construction contract documents?			
32	Whether value engineering study performed, and recommendations taken care?			
33	Whether cost-effectiveness over the entire project life cycle is considered?			
34	Whether designer has considered project boundaries?			
35	Whether the drawings, specifications, and documents prepared to meet TOR requirements?			
36	Whether design deliverables conform to the requirements?			
37	Is there any approved design review procedure that was followed to review and quality check?			
39	Whether all the drawings are numbered?			
40	Whether the reports are complete and include adequate information about the project?			
41	Whether the report is properly formatted and has a table of contents for each report?			
42	Whether drawings were reviewed prior to submission?			
43	Has the designer considered and followed management process activities related to schematic design?			

(*Continued*)

TABLE 8.7 (Continued)
Audit Methodology for Assessment of Schematic Design

Serial Number	Item to be Assessed	Yes	No	Comments
44	Whether design complies with designer's quality management plan?			
45	Whether scope of design work was properly established?			
46	Whether design criteria properly established?			
47	Does the design support proceeding to next design development stage?			
48	Whether the design drawings, reports, and other documents properly labeled and archived?			

F. Interview Questionnaires

Serial Number	Description	Response	Comments
1	Whether there was any delay to collect the data?		
2	Whether there was any delay in site investigation?		
3	Was there any delay in getting authorities' approval?		
4	Whether schematic design prepared within the agreed-upon schedule?		
5	Whether work progress schedule was prepared to monitor the progress?		
6	Whether a timesheet was maintained to record the time spent by each staff engaged for preparation of schematic design?		
7	Whether coordination meetings held among all the team members?		
8	Is there quality organization with responsibilities matrix?		
9	Whether schematic design submitted in time for owner's review?		
10	Whether number of sets prepared as per TOR and submitted?		
11	Whether the owner issued schematic design acknowledgement letter?		
12	Does the schematic design have any major comments from the owner?		
13	Whether the schematic design was not approved by the owner?		
14	Whether the reasons (causes) for comments, non-approval, if any, analyzed, and corrective/preventive action taken?		

TOR lists the requirement guidelines to develop a detailed design. It mainly consists of:

1. Detailed Design Drawings
2. Bill of Quantities
3. Project Specifications
4. Contract Documents
5. Project Schedule
6. Definitive Estimate
7. Format, Scales, and Size of Reports and Drawings

The main objectives of performing a quality audit of the design development phase are:

1. Evaluate and assess process compliance with the organizational strategic policies and procedure.
2. Assess conformance to QMS and processes to meet established quality system requirements.
3. Review the process used to prepare the detailed design.
4. Verify that the detailed design complies with project goals and objectives.
5. Verify that design deliverables are as per TOR requirements.
6. Validate the project schedule and cost of the project meeting the project requirements.
7. Validate compliance with quality standards and codes.
8. Assess compliance with the design quality management system.
9. Check for compliance with regulatory requirements.
10. Check for sustainability considerations in the development of the design.
11. Check that conservation of energy requirements are considered.
12. Check that facility management requirements are considered.
13. Review that key risk factors are analyzed and managed during design development phase for execution of project/facility.
14. Assess compliance with safety consideration in the design.
15. Review design drawings, specifications, and contract documents.
16. Asses that all the major activities are adequately performed and managed.
17. Assess the proper functioning of the project management system.

TABLE 8.8 illustrates the audit methodology for the assessment of the design development phase.

8.5.4 Auditing Process for Construction Documents Phase

Construction documents phase provides a complete set of working drawings of all the disciplines, site plans, technical specifications, Bill of Quantities (BOQ), schedule, (except the standards specifications and documents for insertions normally added during the bidding and tendering phase), and related graphic and written

TABLE 8.8
Audit Methodology for Assessment of Design Development

Serial Number	Item to be Assessed	Yes	No	Comments
G. Checklist				
A-1 Design and Drawings				
1	Whether scope of work for development of detailed design established based on TOR requirements?			
2	Whether organization chart with responsibilities to develop detailed design established?			
3	Whether the detailed design support owner's project goals and objectives?			
4	Whether the detailed design meet all the elements specified in TOR?			
5	Whether the owner's requirements are fully considered?			
6	Whether the design includes all the functional, external works, parking, and landscape areas?			
7	Whether the design comply with organizational strategic policies?			
8	Whether comments on schematic design are taken care of while preparing detailed design?			
9	Whether regulatory/statutory requirements are considered?			
10	Whether the design is based on specified quality standards and codes?			
11	Does the design meet all the performance requirements?			
12	Whether constructability has been taken care?			
13	Whether technical and functional capability is considered?			
14	Is sustainability considered in developing the design?			
15	Whether the design meet LEED requirements?			
16	Is energy conservation considered?			
17	Whether designer has considered facility management requirements?			
18	Whether the designer has taken into consideration project boundaries and assumptions?			
19	Whether value engineering study recommendations are considered while developing the design?			
20	Whether project objectives with respect of time, cost, and quality considered while developing the design?			
21	Does the design support proceeding to next design development stage?			

(Continued)

TABLE 8.8 (Continued)
Audit Methodology for Assessment of Design Development

Serial Number	Item to be Assessed	Yes	No	Comments
A-2 Specifications and Documents				
1	Whether specification prepared taking into consideration international construction contract documents?			
2	Whether the specifications, and documents prepared to meet TOR requirements?			
3	Whether the reports are complete and include adequate information about the project?			
4	Whether the report is properly formatted and has a table of contents for each report?			
5	Whether division numbers in the specifications match with BOQ?			
6	Whether numbers of sets comply with TOR requirements?			
A-3 Schedule				
1	Whether project schedule is practically achievable?			
2	Whether all the activities are considered while preparing the schedule?			
3	Whether activity duration for all the activities are considered while preparing the schedule?			
4	Whether relationship and dependency between activities are identified?			
5	Was schedule developed considering specific internationally recommended schedule-level practices?			
6	Whether detailed design prepared within the agreed-upon schedule			
7	Whether work progress schedule was prepared to monitor the progress?			
8	Whether a timesheet was maintained to record the time spent by each staff engaged for preparation of concept design?			
A-4 Cost				
1	Whether project cost is properly estimated?			
2	Whether cost-effectiveness over the entire project life cycle is considered?			
3	Was cost developed considering specific internationally recommended cost estimation level practices for construction projects?			
A-5 Quality				
1	Whether design is fully coordinated for conflict between different trades?			

(Continued)

TABLE 8.8 (Continued)
Audit Methodology for Assessment of Design Development

Serial Number	Item to be Assessed	Yes	No	Comments
2	Whether designer has considered availability of resources during entire project life cycle?			
3	Whether design complying with designer's quality management plan?			
4	Whether designs were prepared using authenticated and approved software?			
5	Whether all the drawings are numbered?			
6	Whether drawings are reviewed prior to submission?			
7	Whether design was reviewed using mistake-proofing tool to eliminate/mitigate design errors?			
8	Whether design matches with property limits?			
9	Whether legends match the layout?			
10	Whether design drawings are properly numbered?			
11	Whether design drawings have owner logo, designer logo as per standard format?			
12	Whether design calculation sheets are included in the set of documents?			
13	Whether project name and contract reference are shown on the drawing?			
14	Whether BOQ is verified with design drawings and specifications?			
15	Whether detailed design was reviewed as per design review procedure?			
A-6 Risk				
1	Have design risks been identified, analyzed, and responses planned for mitigation?			
2	Whether valid assumptions and constraints considered?			
3	Is there any risk register maintained?			
A-7 Safety				
1	Whether health and safety requirements in the design considered?			
2	Whether environmental constraints and requirements considered?			
3	Whether design confirm with fire and egress requirements?			
4	Whether environmental compatibility considered while developing the design?			
A-8 General				
1	Whether scope of design work was properly established?			

(Continued)

TABLE 8.8 (Continued)
Audit Methodology for Assessment of Design Development

Serial Number	Item to be Assessed	Yes	No	Comments
2	Whether design criteria properly established?			
3	Whether design deliverables are established based on TOR?			
4	Whether all the related data, information collected to develop design?			
5	Whether site investigations done prior to development of detailed design?			
6	Whether design team members are qualified and experienced in developing design of such project?			
7	Whether there is an organization chart for design team?			
8	Whether responsibilities matrix is developed?			
9	Whether regulatory approval obtained and comments, if any, incorporated and all review comments responded?			
10	Has the designer considered and followed management process activities related to the design development?			
11	Whether design is approved by all the concerned stakeholders?			

H. Interview Questionnaires

Serial Number	Description	Response	Comments
1	Whether there was any delay to collect the data?		
2	Whether there was any delay in site investigation?		
3	Was there any delay in getting authorities' approval?		
4	Whether the comments on schematic design were reviewed and taken into consideration?		
5	Was there any approved design submission plan?		
6	Was bill of quantities (BOQ) fully coordinated with design drawings and specifications?		
7	Whether availability of resources during construction phase was explored and considered?		
8	Whether coordination meetings held among all the team members?		
9	Is there quality organization with responsibilities matrix?		
10	Which international contract documents were followed to prepare specifications?		
11	Is there any approved design review procedure?		

(Continued)

TABLE 8.8 (Continued)
Audit Methodology for Assessment of Design Development

Serial Number	Item to be Assessed	Yes	No	Comments
12	Whether detailed design submitted in time for owner's review?			
13	Whether number of sets prepared as per TOR and submitted?			
14	Whether the owner issued detailed design acknowledgement letter?			
15	Does the detailed design have any major comments from the owner?			
16	Whether the detailed design was not approved by the owner?			
17	Whether the reasons (causes) for comments, non-approval, if any, analyzed and corrective/ preventive action taken?			
18	Whether any sub-consultant was involved, if yes, what were the selection and design review criteria?			
19	Whether the design drawings, reports, and other documents properly labeled and archived?			

information to bid the project. It is necessary that utmost care is taken to develop and assemble all the documents and ensure the accuracy and correctness to meet the owner's objectives.

In order to identify requirements to assemble contract documents, the designer has to gather the comments on the submitted detailed design by the owner/project manager, collect TOR requirements and regulatory requirements, identify owner requirements, and all other related information to ensure nothing is missed.

The following is the list of construction documents listed under terms of reference (TOR) that are to be developed by the designer (A/E):

1. Final drawings (working drawings) to be prepared to the required scales, format with necessary logo, client name, location map, north orientation, project name, designer name, drawing title, drawing number, contract reference number, date of drawing, revision number, drawing scale, duly signed by the designer.
2. Bill of quantities and schedule of rates.
3. Contract documents.
4. Technical specifications.
5. Project schedule.
6. Cost estimates.
7. Summary report.

The main objectives of performing a quality audit of the construction documents phase are:

1. Evaluate and assess process compliance with the organizational strategic policies and procedures.
2. Assess conformance to QMS and processes to meet established quality system requirements.
3. Review the process used to prepare construction documents.
4. Verify that the construction documents comply with the requirements for the bidding and tendering process.
5. Verify that construction documents deliverables are as per TOR requirements.
6. Validate the project schedule meeting the project requirements.
7. Validate the cost of the project meeting the project requirements.
8. Validate that working drawings and documents comply with quality standards and codes.
9. Assess compliance of working drawings with the design quality management system.
10. Check for compliance with regulatory requirements.
11. Check for sustainability considerations in the development of design.
12. Check that conservation of energy requirements are considered.
13. Check that facility management requirements are considered.
14. Review that key risk factors are analyzed and managed while preparing construction documents.
15. Assess compliance with safety considerations in the working drawings.
16. Review working drawings, specifications, and contract documents.
17. Asses that all the major activities are adequately performed and managed.
18. Assess the proper functioning of the project management system.

TABLE 8.9 illustrates the audit methodology for the assessment of construction documents.

8.5.5 Auditing Process for Bidding and Tendering Phase

In many countries, it is a legal requirement that government-funded projects employ the competitive bidding method. This requirement gives an opportunity to all qualified contractors to participate in the tender, and normally the contract is awarded to the lowest bidder. Private-funded projects have more flexibility in evaluating the tender proposal. Private owners may adopt the competitive bidding system, or the owner may select a specific contractor and negotiate the contract terms. Negotiated contract systems have flexible pricing arrangements as well as the selection of the contractor based on his/her expertise or the owner's past experience with the contractor

TABLE 8.9
Audit Methodology for Assessment of Construction Documents

Serial Number	Item to be Assessed	Yes	No	Comments
I. Checklist				
1	Whether scope of work for development of construction documents established based on TOR requirements?			
2	Whether organization chart with responsibilities to develop construction documents established?			
3	Whether the construction documents support the owner's project goals and objectives?			
4	Whether the construction documents comply with organizational strategic policies?			
5	Whether the construction documents meet all the elements specified in TOR?			
6	Whether comments on detailed design taken care of while preparing the construction documents?			
7	Whether regulatory/statutory requirements considered?			
8	Whether the design has functional and technical compatibility?			
9	Whether the design is constructible			
10	Do the working drawings meet all the performance requirements?			
11	Whether working drawings meet project objectives in respect of time, cost, and quality?			
12	Were project boundaries considered in the working drawings?			
13	Whether owner's preferred requirements are included in the working drawings?			
14	Whether the design includes all the functional, external works, parking, and landscape areas?			
15	Whether project schedule is practically achievable?			
16	Whether project cost is properly estimated?			
17	Whether BOQ coordinated with Specifications and contract documents?			
18	Whether construction documents phase complying with designer's quality management plan?			
19	Whether the working drawings based on specified quality standards and codes?			
20	Whether working drawings are coordinated with all the trades?			

(Continued)

TABLE 8.9 (Continued)
Audit Methodology for Assessment of Construction Documents

Serial Number	Item to be Assessed	Yes	No	Comments
21	Whether the following information is provided in each of the trade drawings: • Working drawings produced at different scales and formats. • Plans. • Sections. • Elevations. • Schedule. • Drawing index.			
22	Whether all the drawings are numbered?			
23	Whether the following information is included on all the drawings; • Client name. • Client logo. • Location map. • North orientation. • Project name. • Drawing title. • Drawing number. • Date of drawing. • Revision number. • Drawing scale. • Contract reference number. • Signature block. • Signed by the designer for check and approval.			
24	Whether designers have considered availability of resources during construction, testing, and commissioning phases?			
25	Whether recommended material meet owner's objectives?			
26	Whether construction documents conform with the project delivery system?			
27	Whether construction documents conform type of contract/pricing with the adopted methodology?			
28	Whether the following documents are included along with the working drawings: • Existing site. conditions/site plan. • Site surveys. • Design calculations. • Studies and reports.			

(Continued)

TABLE 8.9 (Continued)
Audit Methodology for Assessment of Construction Documents

Serial Number	Item to be Assessed	Yes	No	Comments
29	Whether specification prepared taking into consideration international construction specifications and division numbers?			
30	Whether construction documents prepared taking into consideration international construction contract documents?			
31	Whether risks have been identified, analyzed, and responses planned for mitigation of risks while preparing the working drawings?			
32	Whether health and safety requirements in the working drawings are considered?			
33	Whether design conforms with fire and egress requirements?			
34	Whether environmental constraints considered?			
35	Whether the working drawings, specifications, and documents prepared to meet TOR requirements?			
36	Whether construction documents deliverables conform to the project requirements?			
37	Whether the reports are complete and include adequate information about the project?			
38	Whether the report is properly formatted and has a table of contents for each report?			
39	Is there any approved review procedure?			
40	Whether working drawings are reviewed prior to submission?			
41	Does the management process activities related to prepare construction documents?			
42	Do the working drawings support construction of the project/facility?			
43	Do the documents include penalty for delay in completion of project?			
44	Do the documents include penalty for failing to provide required resources (manpower) for the project?			
45	Do the documents include penalty for delay in violation of safety during execution of project?			
46	Whether the construction documents are properly labeled and archived?			

(Continued)

TABLE 8.9 (Continued)
Audit Methodology for Assessment of Construction Documents

Serial Number	Item to be Assessed	Yes	No	Comments
J. Interview Questionnaires				
Serial Number	**Description**	**Response**		**Comments**
1	What procedure was followed to coordinate between working drawings and specifications?			
2	How the availability of resources during construction phase were ascertained?			
3	Was there any delay in getting authorities' approval?			
4	Whether construction documents prepared within the agreed-upon schedule?			
	Whether work progress schedule was prepared to monitor the progress?			
5	Whether a timesheet was maintained to record the time spent by each staff engaged for preparation of construction documents?			
6	Whether construction documents were submitted in time for owner's review?			
7	What method was used to prepare project schedule?			
8	What methodology was used to estimate definitive cost?			
9	Whether risk factors were considered while developing the construction documents?			
10	Whether number of sets prepared as per TOR and submitted?			
11	Whether the owner issued construction documents acknowledgement letter?			
12	Do the construction documents have any major comments from the owner?			
13	Whether the construction documents were not approved by the owner?			
14	Whether the reasons (causes) for comments, non-approval, if any, analyzed and corrective/ preventive action taken?			

successfully completing one of his/her projects. The negotiated contract systems are based on the following forms of payment:

1. Cost plus contracts: It is a type of contract in which the contractor agrees to do the work for the cost of time and material, plus an agreed-upon amount of profit. The following are the different types of cost-plus contracts:
   a. Cost plus percentage fee contract.

b. Cost plus fixed fee contract.
c. Cost plus incentive fee contract.

2. Reimbursement contracts: It is a type of contract in which the contractor agrees to do the work for the cost per schedule of rates, or bill of quantities, or bill of material.
3. Fixed price contracts: With this type of contract, the contractor agrees to work with a fixed price (also called lump sum) for the specified and contracted work. Any extra work is executed only upon receipt of instruction from the owner. Fixed price contracts are generally inappropriate for work involving major uncertainties, such as work involving new technologies.
4. Target cost contracts: A target cost contract is based on the concept of a top-down approach, which provides a fixed price for an agreed range of out-turn costs around the target. In this type of contract, overrun or underspend are shared by the owner and the contractor at predetermined agreed-upon percentages.
5. Guaranteed maximum price contracts (cost plus guaranteed maximum price): With this type of contract, the owner and contractor agree to a project cost guaranteed by the contractor as maximum.

The bidding and tendering documents are prepared as per the procurement method and contract strategy adopted during the early stage of the project. Tendering procedure documents submitted by the designer are updated and necessary owner-related information is inserted in the tender documents. The bid advertisement material is prepared and upon approval from the owner, the bid notification is announced through different media as per the organization's/agency's policy.

The following stakeholders have direct involvement in the bidding and tendering phase:

- Owner
- Tender Committee
- Designer (Consultant)
- Project Manager/Agency Construction Manager (as applicable)

The main objectives of performing a quality audit for the bidding and tendering process to select a contractor (design-bid-build) for a construction project are:

1. Evaluate and assess process compliance with the organizational strategic policies and procedures.
2. Assess conformance to QMS and processes to meet established quality system requirements.
3. Asses that all the major activities are adequately performed and managed.
4. Check that key risk factors are analyzed and managed to meet business requirements.
5. Assess the proper functioning of the project management system.

In order to manage and control the bidding and tendering process for selection of a contractor can be divided into four major stages. These stages are:

1. First stage: Shortlisting/pre-qualification of designers.
2. Second stage: Tender documents.
3. Third stage: Contract bid solicitation.
4. Fourth stage: Contract award.

FIGURE 8.3 illustrates the bidding and tendering process stages to select a contractor (design-bid-build).

Based on the typical stages discussed above, the audit methodologies can also be performed in four stages. TABLE 8.10 illustrates the audit methodology to asses/evaluate the process compliance at Stage I, TABLE 8.11 illustrates the audit methodology to asses/evaluate the process compliance at Stage II, TABLE 8.12 illustrates the audit methodology to asses/evaluate the process compliance at Stage III, TABLE 8.13 illustrates audit methodology to asses/evaluate the process compliance at Stage IV.

8.5.6 Auditing Process for Construction Phase (Design-Bid-Build Project Delivery System)

Construction is translating the owner's goals and objectives, by the contractor, to build the facility as stipulated in the contract documents, plans, and specifications within budget and on schedule. Construction is the sixth phase of the construction project life cycle and is an important phase in construction projects. A majority of the total project budget and schedule is expended during construction. Similar to costs, the time required to construct the project is much higher than the time required for preceding phases. Construction usually requires a large number of workforce and a variety of activities. Construction activities involve erection, installation, or construction of any part of the project. Construction activities are actually carried out by the contractor's own workforce, or by subcontractors. Construction therefore requires more detailed attention to its planning, organization, monitoring, and control of project schedule, budget, quality, safety, and environmental concerns. The construction phase consists of various activities such as mobilization, submittals, planning and scheduling, management of resources/procurement, execution of work, control and monitoring, quality, and inspection.

The main objectives of performing a quality audit of the construction phase (construction process/works) are to:

1. Determine whether appropriate systems, processes, and controls for managing construction are in place to achieve contractual objectives and owner/client satisfaction.
2. Evaluate and assess process compliance with the organizational strategic policies and procedures.

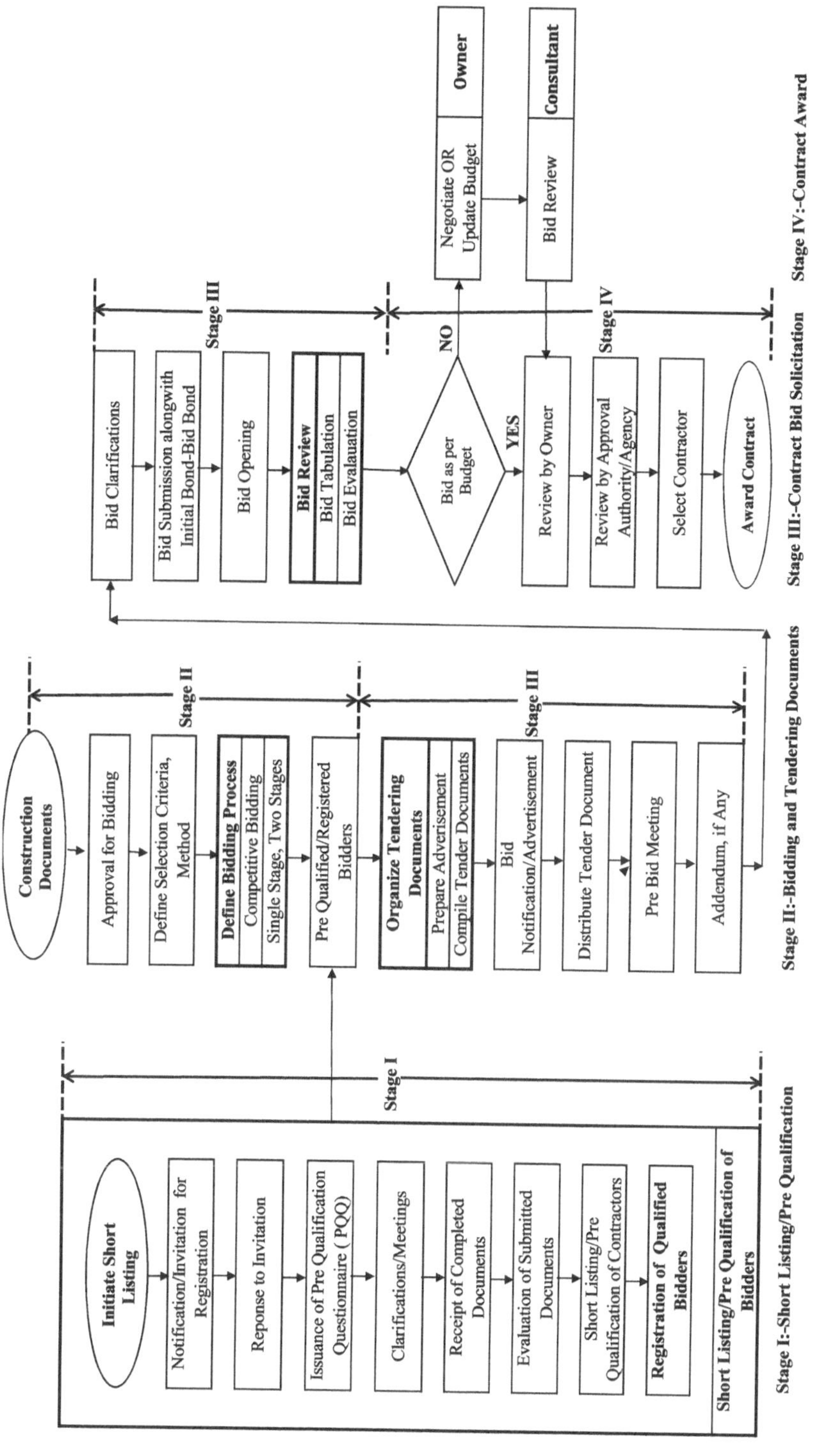

FIGURE 8.3 Bidding and tendering process stages for selection of contractor.

TABLE 8.10
Audit Methodology for Assessment of Bidding and Tendering Stage I

Serial Number	Item to be Assessed	Yes	No	Comments
K. Checklist				
1	Whether the notification for registration was announced in all technical newsletters and leading news media as per the organization's policy?			
2	Whether there are registered designers to participate in the tender?			
3	Whether pre-qualification questionnaires (PQQ) issued to all the intending bidders?			
4	Whether there is any request for queries/bid clarifications?			
5	Whether the meeting for clarification is attended by all the intending bidders?			
6	Whether attendance sheet for the meeting is signed by all the attendees?			
7	Whether minutes of meeting circulated to all the attendees/participating bidders?			
8	Whether bid clarifications to the queries are sent to all the bidders?			
9	Whether sufficient time is provided to prepare and submit completed PQQ?			
10	Is there any request for extension for submission date?			
11	Whether all the intending bidders submitted the response to PQQ?			
12	Whether received bid envelopes were placed in safe custody?			
13	Is there any delayed submission?			
14	Whether received responses acknowledged?			
15	Whether received responses clearly identified and recorded?			
16	Whether responses fairly evaluated as per the organization's selection policy?			
17	Whether evaluation criteria, weightage, and methodology are as specified?			
18	Whether any of the submitted responses found incomplete?			
19	Whether there are any late submissions?			
20	Whether risk factors are considered while evaluating the response and registration of designers?			
21	Whether reasons for nonregistration conveyed to unsuccessful participants?			

(*Continued*)

TABLE 8.10 (Continued)
Audit Methodology for Assessment of Bidding and Tendering Stage I

Serial Number	Item to be Assessed	Yes	No	Comments
22	Whether shortlisted contractors allotted a code and registration number?			
L. Interview Questionnaires				
Serial Number	**Description**	**Response**		**Comments**
1	How responses were submitted: • Hand delivery. • By post. • Courier service.			
2	Whether PQQ was available for electronic distribution?			
3	How was the response for registration/shortlisting?			

TABLE 8.11
Audit Methodology for Assessment of Bidding and Tendering Stage II

Serial Number	Item to be Assessed	Yes	No	Comments
M. Checklist				
1	Whether the tender documents include all the relevant information required to select the contractor (design-bid-build)?			
2	Whether the construction documents approved by the project owner?			
3	Whether bid/tender documents prepared as per the procurement method and contract strategy adopted during early stage of the project?			
4	Whether all the information to enable bidders/contractors to submit the proposal is clear and unambiguous?			
5	Whether documents are prepared taking into consideration organization's quality management system (QMS)?			
6	Are regulatory/authority requirements taken into consideration while preparing the tender documents?			
7	Whether selection criteria and selection method for contractor selection are clearly specified in the organization's QMS?			
8	Whether budget approval obtained for construction prior to release of tender announcement?			
9	Whether tender opening and closing dates mentioned in the announcement?			
10	Is the schedule for completion of construction phase included in the tender documents?			

(Continued)

TABLE 8.11 (Continued)
Audit Methodology for Assessment of Bidding and Tendering Stage II

Serial Number	Item to be Assessed	Yes	No	Comments
11	Whether qualifications of personnel and core team members for construction are requested in the tender documents?			
12	Whether project quality requirements are properly defined in the documents?			
13	Were rights and liabilities of all the parties mentioned in the documents?			
14	Whether change/variation clause included in the documents?			
15	Whether cancellation/termination clauses included in the documents?			
16	Whether bond and insurance clauses are included in the documents?			
17	Whether the tender documents are as per international/local standard documents used by the construction industry?			
18	Whether evaluation and assessment criteria are well defined?			
19	Whether review and analysis procedure of bid is included in the documents?			
20	Whether contract to be signed for construction is included in the documents?			
21	Whether all the records are properly maintained?			
22	Whether bidder's contact details are requested to be included is mentioned in the document?			

N. Interview Questionnaires

Serial Number	Description	Response	Comments
1	Whether any outside agency was involved in preparation of tender documents?		
2	What guidelines were adopted to prepare the tender documents?		
3	Whether any specific format/template was followed?		
4	Whether all the requirements of tender documents were coordinated and agreed by relevant stakeholders?		
5	Whether risks of disputes and associated financial costs considered while preparing the contract documents?		
6	Whether problems and challenges identified while preparing bid/tender?		
7	Who was responsible to authorize bidding requirements?		

TABLE 8.12
Audit Methodology for Assessment of Bidding and Tendering Stage III

Serial Number	Item to be Assessed	Yes	No	Comments
O. Checklist				
1	Whether tender documents properly organized?			
2	Whether the tender documents are distributed to all the prequalified/short-listed/registered bidders (contractors) to participate in the tender?			
3	Whether tender is notified to all the prequalified bidders?			
4	Whether all the prequalified/short-listed/registered bidders participate in the tender?			
5	Whether attendance sheet for the meeting was signed by all the attendees?			
6	Whether minutes of meeting circulated to all the attendees/participating bidders/contractors?			
7	Whether sufficient time is provided to prepare and submit the tender?			
8	Whether tender submission date is extended from the originally announced date?			
9	Whether addendum, if any, are issued/notified to all the participating bidders?			
10	Is there any request for extension for submission date?			
11	Whether clarification meeting attended by all participating bidders?			
12	Whether any changes made to the originally announced tender documents?			
13	Whether received bid envelopes placed in safe custody?			
14	Whether technical and financial envelopes submitted at the same time?			
15	Is there any delayed submission?			
16	Whether tender opened as per announced date and time?			
17	Whether tender opened in presence of all the bidders?			
18	Whether received tenders acknowledged?			
19	Whether received tenders are clearly identified and recorded?			
20	Whether applicable fees/bid bonds submitted by all the bidders who participated in the tender?			

(*Continued*)

TABLE 8.12 (Continued)
Audit Methodology for Assessment of Bidding and Tendering Stage III

Serial Number	Item to be Assessed	Yes	No	Comments
21	Whether tenders are fairly evaluated as per the announced policy and procedure?			
22	Whether comparison of bids are tabulated?			
23	Whether evaluation criteria, weightage, and methodology are as specified?			
24	Whether bid cost is higher than the budgeted cost?			
25	Whether any submitted tender is found incomplete?			
26	Whether there is any late submission?			
27	Whether all the scheduled pages are signed by the bidder?			
28	Whether risk factors are considered while evaluating the bids?			
29	Whether reasons for non-acceptance of tender was conveyed to the unsuccessful bidders?			

P. Interview Questionnaires

Serial Number	Description	Response	Comments
1	How was the notification announced?		
2	How tenders were submitted: • Hand delivery. • By post. • Courier service.		
3	Whether tender was available for electronic distribution?		
4	How was the response to participate in tender?		
5	Whether tender box(es) was placed at secured place and monitored through electronic surveillance system?		
6	How much time was allowed to open the bids after the notified tendering closing time?		
7	Whether tender submission date was extended and what was the reason for extension?		
8	Whether all the tender envelopes/packages were sealed with marking of tenders and contain original tender documents and required number of sets as per tender conditions?		
9	Whether bid review was established?		
10	Whether bid selection procedure was clearly defined?		

TABLE 8.13
Audit Methodology for Assessment of Bidding and Tendering Stage IV

Serial Number	Item to be Assessed	Yes	No	Comments
Q. Checklist				
1	Whether the selection of contractor satisfy all the conditions to be a successful bidder?			
2	Whether the selected bidder (contractor) is a legal entity?			
3	Whether the selection was made by the selection committee?			
4	Whether the selection of contractor approved by relevant applicable authority if mandated by the organization's policy?			
5	Whether the selected contractor is capable of carrying out the specified work and has the resources?			
6	Whether risk factors are considered prior to signing of contract?			
7	Whether standard contract format used for signing the contract?			
8	Are contract terms and conditions clearly written and unambiguous?			
9	Whether addendum and minutes of meeting included in the contract documents?			
10	Whether contract period/schedule properly described?			
11	Whether staff qualifications (core staff) properly defined?			
12	Whether performance bond submitted by the consultant?			
13	Whether variation/change management clauses included in the contract agreement?			
14	Whether dispute resolution, conflict resolution clauses properly defined?			
15	Whether contract documents reviewed prior to signing?			
16	Whether tender documents and contract properly archived?			
17	Whether letter of award is sent?			
18	Whether final results published and announced?			

R. Interview Questionnaires

Serial Number	Description	Response	Comments
1	On what basis the bidder (contractor) was selected: low bid, competitive bid, or quality-based system?		

(*Continued*)

TABLE 8.13 (Continued)
Audit Methodology for Assessment of Bidding and Tendering Stage IV

Serial Number	Item to be Assessed	Yes	No	Comments
2	Was there any tie between two or more bidders?			
3	Were any disputes or objections raised by any of the bidders before award of contract to the successful bidder?			
4	Whether final price was negotiated or as per the bid (quotation) received?			
5	Whether successful bidder failed to submit performance bond?			
6	Whether there was any need to update the approved budget?			
7	Whether contractor declared any subcontractor for outsourcing (subcontracting) purposes?			
8	Whether risks in project execution was considered?			
9	Whether the contract includes penalty clause for delay in completion of project?			
10	Whether the contract includes penalty clause for failure to provide required resources (manpower) project?			
11	Whether the contract includes penalty clause for safety violations during execution of the project?			

3. Assess conformance to corporate QMS and processes to meet project-specific quality system requirements.
4. Review the process used for the construction/execution of work.
5. Ensure that the construction of the project complies with the contract requirements.
6. Ensure the responsibilities matrix is properly established taking into consideration all the stakeholders.
7. Validate to what extent project execution requirements (scope of work) are identified and implemented.
8. Validate project schedule and cost is properly managed and controlled.
9. Validate compliance with quality standards and codes.
10. Assess compliance with the contractor's quality control plan.
11. Check for compliance with regulatory requirements.
12. Verify the extent to which the resources are utilized.
13. Verify that the selection of material/system complies with contract requirements and the organization's procurement policy.
14. Verify that all the submittals are as per the approved schedule.
15. Verify that project documents and logs are maintained.

16. Review that key risk factors are analyzed and managed during the construction phase for the execution of the project/facility as per the risk management plan.
17. Assess compliance with safety considerations as per the safety management plan.
18. Review design drawings, specifications, and contract documents.
19. Asses that all the major activities are adequately performed and managed.
20. Assess the proper functioning of the project/construction management system.
21. Verify the extent to which the project construction complied with the rules and principles of the project/construction management system.
22. Verify the selection of subcontractors complies with contract documents and the organization's policy.

TABLE 8.14 illustrates audit methodology for the assessment of construction/execution activities/processes during the construction phase.

TABLE 8.14
Audit Methodology for Assessment of Construction Phase

Serial Number	Item to be Assessed	Yes	No	Comments
S. Checklist				
A-1 Construction/Execution of Works				
1	Whether project execution requirements are properly established?			
2	Whether the shop drawings approved?			
3	Whether the materials/systems approved?			
4	Whether method of installation (method statement) approved?			
5	Whether execution/installation is carried out as per approved/recommended processes?			
B-1 Monitoring and Controlling				
1	Whether project monitoring and control process as per project requirements?			
2	Whether change management system is in place?			
3	Whether project progress is monitored by traditional monitoring system or by digitized monitoring system?			
4	Whether critical activities that may affect the project are identified?			
5	Whether progress photos are taken to monitor work progress?			
B–1.1 Stakeholders				
1	Whether project stakeholders identified and listed?			
1	Whether responsibilities matrix of all the stakeholders is established?			

(Continued)

TABLE 8.14 (Continued)
Audit Methodology for Assessment of Construction Phase

Serial Number	Item to be Assessed	Yes	No	Comments
2	Whether reports are distributed to all the stakeholders?			
3	Whether meeting invitation is sent to all the stakeholders?			
4	Whether core staff is approved and available at the project?			
5	Whether all the approvals from the authorities obtained?			
B–1.2 Scope				
1	Whether work is executed as per defined scope and agreed methodology?			
2	Is there any change in the contracted scope of work and specifications?			
3	Whether preventive and corrective measures are being taken to control the scope?			
4	Is there any request for information (RFI) raised?			
5	Is there any request for variation submitted?			
6	Is there any request for alternate material?			
7	Is there any site work instruction (SWI)?			
B–1.3 Schedule				
1	Whether contractor's construction schedule is approved?			
2	Whether sequence of activities properly coordinated?			
3	Whether performance baseline for all the activities identified?			
4	Whether all the submittals are as per approved schedule and monitored?			
5	Whether project progress status is prepared to know the actual progress versus planned progress?			
6	Whether project forecasting is done to predict future progress of activities to be performed?			
7	Whether percentage of completion of each activity is based on approved checklist?			
8	Whether schedule is updated based on approved changes?			
9	Whether subcontractor's work progress is monitored?			
10	Whether contractor was submitting daily progress report?			

(Continued)

TABLE 8.14 (Continued)
Audit Methodology for Assessment of Construction Phase

Serial Number	Item to be Assessed	Yes	No	Comments
11	Whether contractor was submitting daily manpower, equipment reports available at site?			
B–1.4 Cost				
1	Whether there is an approved construction budget?			
2	Whether there is approved S-Curve?			
3	Whether cash flow is properly maintained?			
4	Whether earned value management in place?			
5	Whether the project cost is exceeding the planned budget?			
6	Whether forecasted cost is as planned or likely to exceed?			
7	Whether there is any approved variation cost to the contracted project cost?			
8	Whether S-Curve is updated based on approved variation orders?			
B–1.5 Quality				
1	Whether contractor's quality control plan is approved?			
2	Whether builder's workshop drawings, coordinated drawings, composite drawings, shop drawings are approved?			
3	Are there any comments on the shop drawing approval transmittals?			
4	Whether all the materials/systems approved?			
5	Are there any comments on the submitted material transmittals?			
6	Whether incoming material is inspected upon receiving at site?			
7	Whether test certificates, country of origin submitted along with the material incoming checklist?			
8	Is there any rejection to the material received at site?			
9	Is there any factory inspection carried out for the material/system?			
10	Whether work is being carried out as per approved builder's workshop drawings, coordinated drawings, composite drawings, and shop drawings?			
11	Whether work is carried out as per approved methods?			

(Continued)

TABLE 8.14 (Continued)
Audit Methodology for Assessment of Construction Phase

Serial Number	Item to be Assessed	Yes	No	Comments
12	Whether approved materials/systems are installed?			
13	Is there any variation to the approved methods?			
14	Whether quality meetings are conducted?			
15	Whether works are executed to comply with regulatory requirements?			
16	Whether subcontractor's quality management system is properly monitored to comply with contract documents?			
B–1.6 Resources				
1	Whether project staff and other workforce requirements were established?			
2	Whether project staff and workforce are available at site as per the approved manpower/staff deployment chart?			
3	Is there any conflict among project team members?			
4	Whether sufficient workforce available to meet the project execution plan?			
	Whether procurement log is properly maintained?			
5	Whether equipment, machinery available at site to execute the construction works?			
6	Whether subcontractors have sufficient resources to execute the work as per schedule?			
7	Whether work is assigned on daily basis?			
8	Whether attendance sheet is maintained?			
9	Whether subcontractor management is in place?			
B–1.7 Communication				
1	Whether site administration matrix is in place?			
2	Whether logs are maintained?			
3	Whether communication plan is in place?			
4	Whether correspondence procedure between all the stakeholders is established?			
5	Whether communication method is established?			
6	Whether not approved transmittals are addressed promptly?			
7	Whether logs are updated on regular basis?			
8	Whether progress meetings, coordination meetings are held as scheduled?			
9	What is the frequency of progress meetings, quality meetings, HSE meetings?			
10	Whether daily, weekly, and monthly reports are properly prepared and circulated to the relevant stakeholders?			
11	Whether project documents are kept in safe and accessible location?			

(Continued)

TABLE 8.14 (Continued)
Audit Methodology for Assessment of Construction Phase

Serial Number	Item to be Assessed	Yes	No	Comments
B–1.8 Risk				
1	Whether risk management plan is in place?			
2	Whether risk register is maintained and updated?			
3	Whether occurrence of new risk is reported to update project progress and performance?			
4	Whether risk response plan is prepared to address the identified and assessed risks?			
5	Whether effectiveness of risk response plan is regularly evaluated, and effective measures taken?			
6	Whether financial risk is reported to the owner/client?			
B–1.9 Contract				
1	Whether contract management process is in place?			
2	Whether project procurement requirements are identified?			
3	Whether bidding and tendering procedure is followed to procure the material, system, machinery, and equipment?			
4	Whether subcontractors list is finalized?			
5	Whether procurement follow-up system is in place?			
6	Whether cost of subcontracted items is as per estimated value?			
7	Whether there is any delay in receipt of material, system, equipment that may affect project execution schedule?			
8	Whether subcontractor's works are as per schedule?			
B–1.10 HSE				
1	Whether HSE management plan is approved?			
2	Whether safety meetings are conducted?			
3	Whether safety drills are conducted?			
4	Whether temporary firefighting system is operative?			
5	Whether accident preventive measures are followed?			
6	Whether first aid and emergency medical services are available at site?			
7	Whether accident reporting system is followed properly?			
8	Whether accident records are maintained?			
9	Whether hazardous area is marked?			
10	Whether personnel are using protective equipment and other safety gear?			

(Continued)

TABLE 8.14 (Continued)
Audit Methodology for Assessment of Construction Phase

Serial Number	Item to be Assessed	Yes	No	Comments
11	Whether periodic safety inspection is carried out and what is the frequency of such inspection?			
12	Whether safety violation report is issued?			
13	Whether safety disciplinary notice is issued?			
14	Whether access and escape route signs are installed at site?			
15	Whether machinery, equipment, and tools are regularly inspected for safe usage?			
16	Whether crane and hoist have valid third-party certification?			
17	Whether protective clothing and equipment are used while concreting?			
18	Whether sirens and bells are in working conditions?			
19	Whether barricade is installed to avoid accidents?			
20	Whether all formwork and scaffoldings are checked as per permissible load prior to casting?			
B–1.11 Finance				
1	Whether there is contractor's finance plan in place?			
2	Whether progress payment is submitted on time?			
3	Whether cash flow is as per approved S-Curve?			
4	Whether funding limit reconciliation is done periodically?			
5	Whether the project progress payment is received on time as per contract?			
6	Whether there is any deduction from the approved progress payment?			
7	Whether subcontractors are paid as per contract?			
8	Is there interim payment, advance payment for procurement of material?			
9	Whether staff salaries are paid on regular basis?			
10	Whether payment to the suppliers is as per the agreed upon terms?			
11	Whether there is any funding agency to finance the project?			
12	Whether finance records and logs are maintained properly?			
B–1.12 Claim				
1	Whether any claim is identified, recorded, and informed to the client/owner?			
2	Whether there is any dispute noticed related to contract documents?			
3	Is there any claim that may affect project progress?			

(Continued)

TABLE 8.14 (Continued)
Audit Methodology for Assessment of Construction Phase

Serial Number	Item to be Assessed	Yes	No	Comments
C. Interview Questionnaires				

Serial Number	Description	Response	Comments
1	Whether the project team members know the scope of work to executed?		
2	Whether the project team members know the construction phase deliverables?		
3	Whether all the related data of each activity are collected and recorded for project monitoring purpose?		
4	Whether there is an organized and efficient means for measuring, collecting, verifying, and quantifying data to monitor progress of the project?		
5	Whether project deliverables are as defined in the scope baseline or there are any changes in the scope baseline?		
6	Whether there is any approved procedure for integrating the changes?		
7	Was there any system to initiate the changes/ variation and if there are any standard forms?		
8	What tools and techniques are used to improve schedule and recover delays?		
9	What is the frequency of progress status reporting?		
10	Whether earned value management (EVM) methodology is used to measure and evaluate project performance?		
11	Whether there are not approved checklists and how rework is managed?		
12	Whether productivity rate and performance of workers are as estimated during tendering stage?		
13	Whether material management process is properly followed?		
14	Whether identified risks are used to update project schedule and cost?		
14	Whether risk management process cycle is properly followed?		
15	Whether risk probability level is being followed to determine risk impact on the project?		
16	What is the procedure followed to select the subcontractor?		
17	Whether there are any items expensive than the estimated cost?		
18	Whether supply chain management system is followed while selecting the supplier?		
19	Whether safety awareness/training programs are conducted?		

(Continued)

TABLE 8.14 (Continued)
Audit Methodology for Assessment of Construction Phase

Serial Number	Item to be Assessed	Yes	No	Comments
20	Whether emergency evacuation plan is prepared?			
21	Whether evacuation routes are displayed?			
22	What method is followed to resolve the claim?			
23	Whether the reasons (causes) for comments on shop drawings, if any, analyzed and corrective/ preventive action taken?			
24	Whether the reasons (causes) for rejection of material, if any, analyzed and corrective/ preventive action taken?			
25	Whether the reasons (causes) for rejection/ non-approval of executed works, if any, analyzed and corrective/preventive action taken?			

8.6 RISKS IN AUDITING

The following are the risks in quality auditing:

- The scope and objective of audits are not clearly defined.
- The auditing person/agency does not have expertise or knowledge about the type of audit to be performed.
- Auditor does not attempt or cannot check all the transactions thatneed to be audited.
- Auditor does not identify and assess the risk.
- Auditing procedure does not cover all the broad areas.
- Auditor's understanding of the entity and the environment including internal control systems in which it operates.
- Assessment of risk.

9 Value Methodology

Abdul Razzak Rumane

9.1 VALUE METHODOLOGY

9.1.1 Value

Value is difficult to define because it is used in a variety of ways. There are seven classes of value that are recognized today. These are:

1. Economic
2. Moral
3. Aesthetic
4. Social
5. Political
6. Religious
7. Judicial

Of all these classes, only economic value can be measured in terms of (hopefully) objective monetary units such as dollars, euros, yen, or dinars. However, economic value is also established through an item's use value (facilities that provide a unit of use work, or service) and esteem value (properties that make something desirable). We can say that use values cause a product to perform, and esteem values cause it to sell. Use value and esteem value defies precise quantification in monetary terms so we often resort to multi-attribute techniques for evaluating the total value of complex designs and complicated systems or machinery. Accordingly, value can be defined as:

Value = Quality / Cost or Value is quality divided by cost.

It means that if you can provide better facilities to the client at a lower price, the value goes up.

9.1.2 Value Engineering

Value Engineering (VE) is a technique for evaluating the design of a product/project to ensure that essential functions are provided at minimal overall cost to the manufacturer or client/end user. It is a systematic approach to identifying the project's functional objectives with the goal of optimizing design, construction, and future operations. VE is a creative approach to optimizing value. The authors of *Quality in Constructed Projects* by ASCE have described that:

DOI: 10.1201/9781003636687-9

In its broadest sense, project value is determined by the relationship of the worth of the project and its elements to their cost. The following expression summarizes the relationship:

$$\text{Value} \sim (\text{Function} + \text{Performance} + \text{Quality}) / (\text{Cost})$$

The goal of the VE process is to identify alternatives that maximize this value relationship. The owner's requirements are the basis for establishing values for the items in the numerator. For example, an owner who intends to sell a facility within five years of its completion may place less value on long-term maintenance characteristics than an owner who intends to keep a facility for decades. Owner requirements directly affect the relative value of aesthetics, reliability, sustainable development, maintainability, operability, construction duration, and other characteristics. Improving the quality of elements related to these characteristics usually increases costs. The goal of VE is to achieve a ratio of quality to cost that is acceptable to and in the best interest of the owner.

9.1.2.1 Objective of Value Engineering

The objective of VE is to provide the required product/project functions at a minimum cost and to achieve equivalent or better performance while maintaining all functional requirements defined by the customer/client. VE necessitates a detailed examination of a product/project function and the cost of each item in addition to a thorough review of product/project specifications. It does this by identifying and eliminating unnecessary costs.

9.1.2.2 Value Engineering Team

VE studies are conducted by a multidisciplinary team that focuses on a clearly defined scope. The VE team consists of specialists from a variety of disciplines (design, manufacturing, marketing, etc., for product evaluation; and owner, design professional, contractor, consultant, and construction management/project management, etc., for construction project) focusing on determining the most effective way to provide high value at an accepted cost to the customer/client end user. While each member of the project team is free to recommend that a VE study be undertaken, it is typically the owner's responsibility to authorize and formally initiate the VE effort.

The success of a VE effort is linked to the experience of the team members. The number and qualifications of VE team members depend on project objectives and the stage of the project life cycle at which VE studies are conducted. VE studies typically involve multiple disciplines. VE team members must be from diverse disciplines and backgrounds and have a range of expertise related to the project's key issues. Each of the team members must have:

- Positive attitude and approach.
- Technical knowledge related to the product/project.
- Higher education.
- Professional experience.
- Certification from a recognized body as an expert in VE.

The VE team is led by a Certified Value Specialist (CVS). SAVE International (The Value Society) manages the certification program for CVS.

9.1.2.3 Approach

The VE approach involves brainstorming such questions as:

- Does the item have any design features that are not necessary?
- Can two or more parts be combined together?
- How can we cut down the weight?
- Are these nonstandard parts that can be eliminated?

The key to successful VE is to ask critical questions and seek creative answers. The following are sample questions that should be included in a VE study:

- Are all the functions provided required by the customer?
- Can less expensive material be used?
- Can the number of different materials used be reduced?
- Can the design be simplified to reduce the type of items?
- Can a product designed for another project be used?
- Are all surfaces and finishes necessary?
- Would product redesign eliminate a quality problem?
- What is the level of technology required to achieve the owner's objectives?

It is important to question everything and not take anything for granted. Creative solutions can be obtained using classical brainstorming or the nominal group technique.

9.1.2.3.1 Classical Brainstorming

Classical brainstorming is the most well-known and often-used technique for idea generation in a short period of time. It is used when creative and original ideas are desired and is based on the fundamental principles of deferment of judgment and that quantity breeds quality. The following are the rules for successful brainstorming:

1. Criticism is ruled out—no criticism.
2. Freewheeling is welcomed.
3. Quantity is wanted—more ideas are better.
4. Contribution and improvement are sought participants should speak whatever ideas pop into their minds.
5. Avoid evaluation—no comments on the idea.

A classical brainstorming session has the following basic steps:

1. Preparation: During this period, the participants are selected, and a preliminary statement of the problem is circulated among them.
2. Brainstorming: A warm-up session with simple unrelated problems is conducted, the relevant problem and the rules of brainstorming are presented,

and ideas are generated and recorded using agreed-upon checklists/formats and other techniques if necessary.
3. Evaluation: The ideas are evaluated relative to the problem.

Generally, a brainstorming group consists of four to seven people, although some suggest a larger group.

9.1.2.3.2 Nominal Group Technique

The Nominal Group Technique (NGT) involves a structural group meeting designed to incorporate individual ideas and judgments into a group consensus. By correctly applying the NGT, it is possible for groups of people (preferably five to ten) to generate alternatives or other ideas for improving the competitiveness of the product/project.

The NGT, when properly applied, draws on the creativity of the individual participants while reducing undesirable effects of most group meetings such as:

- The dominance of one or more participants.
- The suppression of conflict ideas.

The basic format of an NGT session is as follows:

1. Individual silent generation of ideas.
2. Individual round-robin feedback and recording the ideas.
3. Group's clarification of each idea.
4. Individual voting and ranking to prioritize ideas.
5. Discussion of group consensus results.

The NGT session begins with an explanation of the procedure and a statement of questions, preferably written by the facilitator. The group members are then asked to prepare an individual list of alternatives that they feel are crucial to the survival and health of the organization.

After this phase has been completed, the facilitator calls on each participant, in a round-robin fashion, to present one idea from his/her list. Each idea (or opportunity) is then identified in turn and recorded on a flipchart or board by the NGT facilitator, leaving sufficient time between ideas for comments or classification. This process continues until all the opportunities have been recorded, clarified, and displayed for all the members to see. At this point, a voting procedure is used to prioritize ideas or opportunities. Finally, voting results lead to the development of group consensus on the topic being addressed.

9.1.2.4 Timings of Value Engineering

VE studies benefit project quality at most stages of the project phases. Conducted in the early stages of a project, they tend to provide the greatest benefit, as there is more potential for cost saving. FIGURE 9.1 illustrates the effectiveness of VE. In fact, in recent years, VE has become an aid to owners in the formulation of goals and objectives before most design work begins.

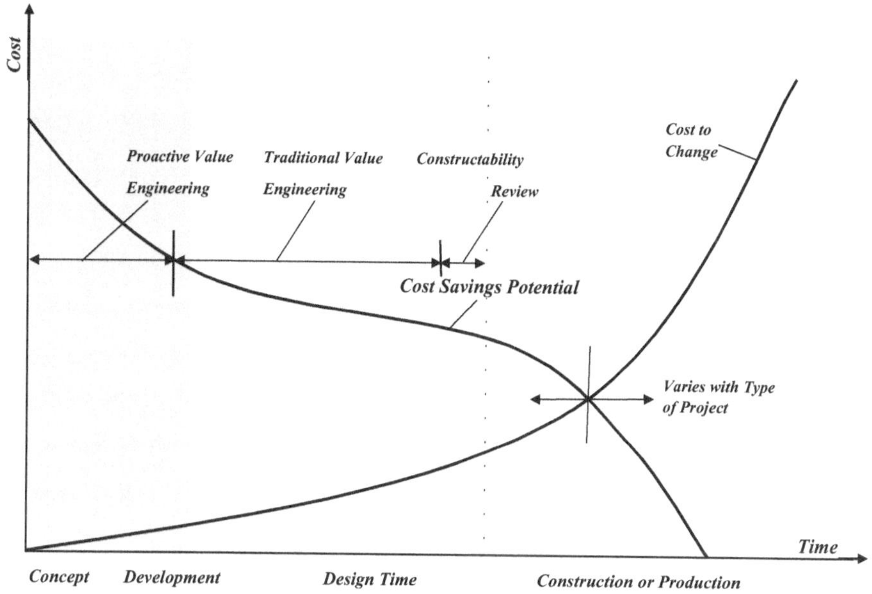

FIGURE 9.1 Effectiveness of value engineering.

VE studies for construction projects are conducted at one or more of the following project stages:

- Conclusion of concept design or facilities planning.
- 30% design completion (preliminary design).
- 60% design completion.
- 90% or 100% design completion (detailed design).

VE enhances project quality at every stage of the project phase. In construction projects, VE starts during the conceptual phase.

Concept design: VE performed during the conceptual design phase improves the quality of project design. During this phase, VE involves the investigation of alternatives and design concepts with the client/end user, design professional, project or construction manager, and any other project team member involved in this phase. The project team works to identify the client/end user's requirements and to design concepts with the potential to achieve the owner's objectives.

Preliminary design: Major changes resulting from VE studies conducted during this phase can be implemented without significant impacts to the project schedule or budget. The potential for VE-Related cost savings is quite substantial during this phase. VE studies conducted during this phase conform to the owner's/end user's goals and objectives.

Detailed engineering: VE studies conducted during this phase (60% completion or more) focus on detailed design and construction methods and the constructability of the project. TABLE 9.1 illustrates areas of VE study by design stage.

TABLE 9.1
Area of Value Study by Design Stage

	Area of Value Study		
Area of Study	**Conceptual**	**Schematic (Preliminary)**	**Design Development (Detailed Design)**
General Project Budget Layout Criteria and Standards	• Design Concepts • Program Interpretation • Site/Facility Massing • Access, Circulation • Design Intentions • Net to Gross Ratios	• Schematic Floor Plans • Schematic Sections • Approach to Systems Integration • Floor to Floor Height • Functional Space	• Floor Plans • Sections • Typical Details • Integrated Systems • Space Circulation • Specifications
Structural Foundation Substructure Superstructure	• Performance Requirements • Structural Bay Sizing • Framing Systems Exploration • Sub-Surface Conditions • Underground Concepts • Initial Framing Review • Structural Load Criteria	• Schematic Basement Plan • Selection of Foundation Systems • Structural System Selection • Framing Plan Outline • Sizing of Elements	• Basement Floor Plan • Key Foundation Elements Details • Floor and Roof Framing Plans • Sizing of Major Elements • Outline Specifications
Architectural Exterior Closure Roofing Interior Construction Elevators Equipment	• Approach to Elevation Views to/from Building • Roof Types and Pitch • Interior Design • Configuration of Key Rooms • Organization of Circulation Schemes • Need and Type of Vertical Circulation • Impact of Key equipment on Facility and Site • Passive Solar Usage	• Concept Elaboration • Selection of Wall Systems • Schematic Elevations • Selection of Partitions • Circulation Sizing • Basic Elevator and Vertical Transportation Concepts • Impact of Key Equipment on Room Design	• Elevations • Key Elevation Details • Key Roofing Details • Initial Finish Schedule • Interior Construction Elements • Integration of Structural Framing • Key Interior Elevations • Equipment Items

(Continued)

TABLE 9.1 (Continued)
Area of Value Study by Design Stage

Area of Study	Area of Value Study: Conceptual	Area of Value Study: Schematic (Preliminary)	Area of Value Study: Design Development (Detailed Design)
Mechanical HVAC Plumbing Fire Protection	• Basic Energy Concepts • Impact of Mechanical Concepts on Facility • Initial Systems Selection • Source Allocation • Performance Requirements for Plumbing, HVAC, Fire Protection	• Mechanical Systems Selection • Refinement of Services and Distribution Concepts • Input to Schematic Plans • Energy Conservation	• Detailed System Selection • Initial system Drawings and Key Details • Distribution and Riser Diagrams • Outline Specifications for System Design
Electrical Service and Distribution Lighting and Power	• Basic Power Supply • Approach to use of Natural and Artificial Lighting • Performance Requirements for Lighting • Need for Special Electrical Systems	• Windows/Skylight Design and Sizing • Selection of Lighting and Electrical Systems • General Service, Power and Distribution Concepts	• Detailed Systems Selection • Distribution Diagrams • Key Space • Lighting Layouts • Outline Specifications for Electrical Elements
Site Preparation Utilities Landscaping	• Site Selection • Site Development Criteria • Site Forms and Massing • Requirements for Access • Views to/from Facility • Utility Supply • Site Drawings	• Design Concept Elaboration • Initial Site Plan • Schematic Planting, Grading, Paving Plans	• Site Plan • Planting Plan • Typical Site Details • Outline Specifications for Site Materials

Source: Dell'Isola, Alphonse. (1997). *Value Engineering: Practical Applications*, reprinted with permission of RSMeans, Gordian and Wiley.

Construction: VE recommendations made during the construction phase are called Value Engineering Change Proposals (VECPs). These usually relate to alternatives/substitutes of material, equipment, and construction methods normally proposed by the contractor. Savings made due to VE application during this phase are adjusted per agreed-upon terms of the contract.

Target cost is the amount the owner would like to spend to build the facility. In construction projects, the target-based approach can be applied to the development of the facility from the investment point of view, having a fixed budget allocation. The design is reviewed, keeping in mind the budget price of the facility to ensure that the facility shall be functionally qualitative and constructed within the targeted budget. If the total estimated cost is greater than the target cost, then the design must be fed back into the VE process.

9.1.2.5 Stages of Value Engineering Study

FIGURE 9.2 illustrates the VE process summary.

VE studies typically consist of three sequential stages:

1. Preparation
2. Workshop (execution of the VE job plan)
3. Post-Workshop

Preparation: This is considered by many to be the most important step of the VE study. During this phase, the owner's need is defined, VE team members are selected, they are informed about the project, and the scope of the study is established.

Workshop: The VE team convenes a workshop that normally has a six-phase job plan. These are:

1. Gathering Information
2. Functional Analysis

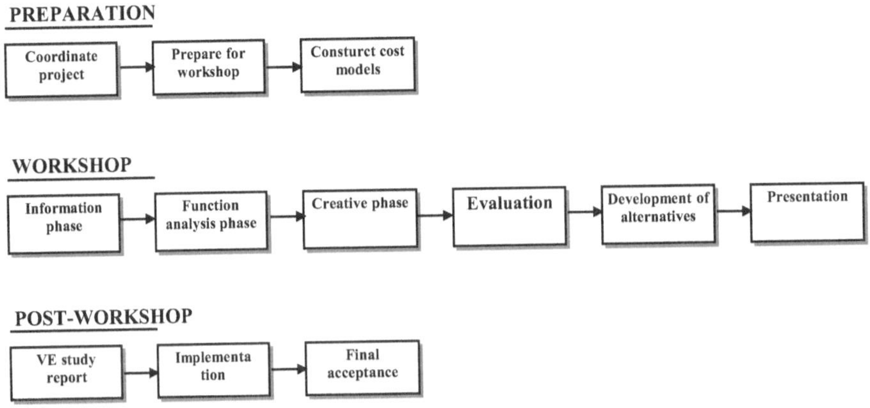

FIGURE 9.2 Value engineering process summary.

3. Creating Alternatives
4. Evaluation
5. Development
6. Presentation

Post-Workshop: After the workshop, the VE team prepares a report to support its findings and the implementation program. Their recommendations are reviewed by the owner and the design team before implementation.

FIGURE 9.3 illustrates the activities performed during the VE study process.

9.1.2.6 Function Analysis System Technique (FAST)

FAST is a technique that specifically illustrates the relationship of all functions within a specific project utilizing a "how-why" logic pattern based on intuitive logic. It is a helpful diagnostic tool, asking questions that generate basic information about project functions. The FAST diagram is especially helpful if the VE team is uncertain of project goals. The following are types of FAST diagrams: Classic, customer/user, and technical. FIGURE 9.4 illustrates a typical FAST concept diagram.

9.1.2.7 The Benefits of Value Engineering

The benefits of VE usually come from improvements in the efficiency of the project delivery system, refinements to specific features, or the development of new approaches to achieving the owner's requirements. The first item most owners look at when they consider VE is potential cost savings. However, VE focuses on improving the relationship of function, performance, and quality to cost—not merely cutting costs.

The benefits of VE in construction are:

- Reduce project construction costs.
- Improve project schedule.
- Simplify procedures.
- Increase procurement efficiency.
- Use resources more effectively.
- Decrease operation and maintenance costs.

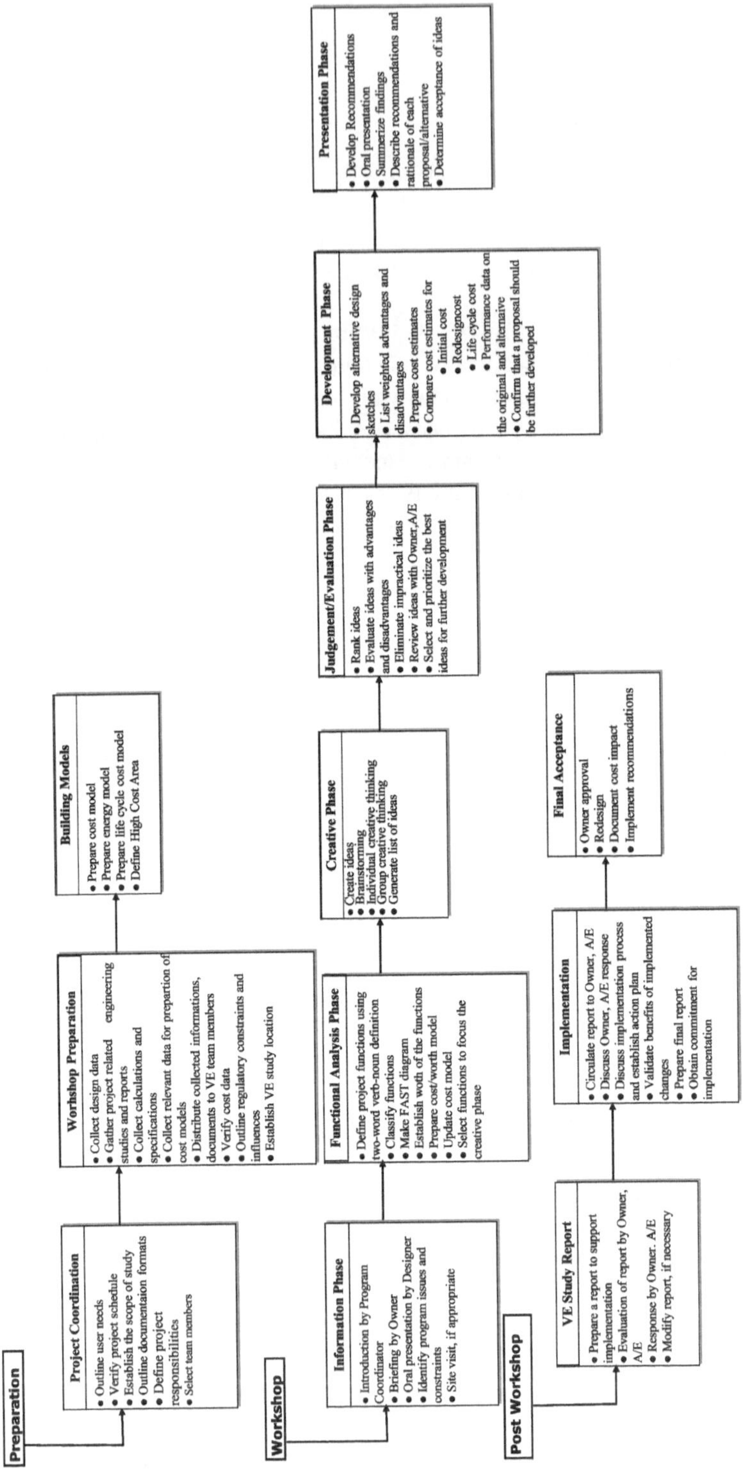

FIGURE 9.3 VE study process activities.

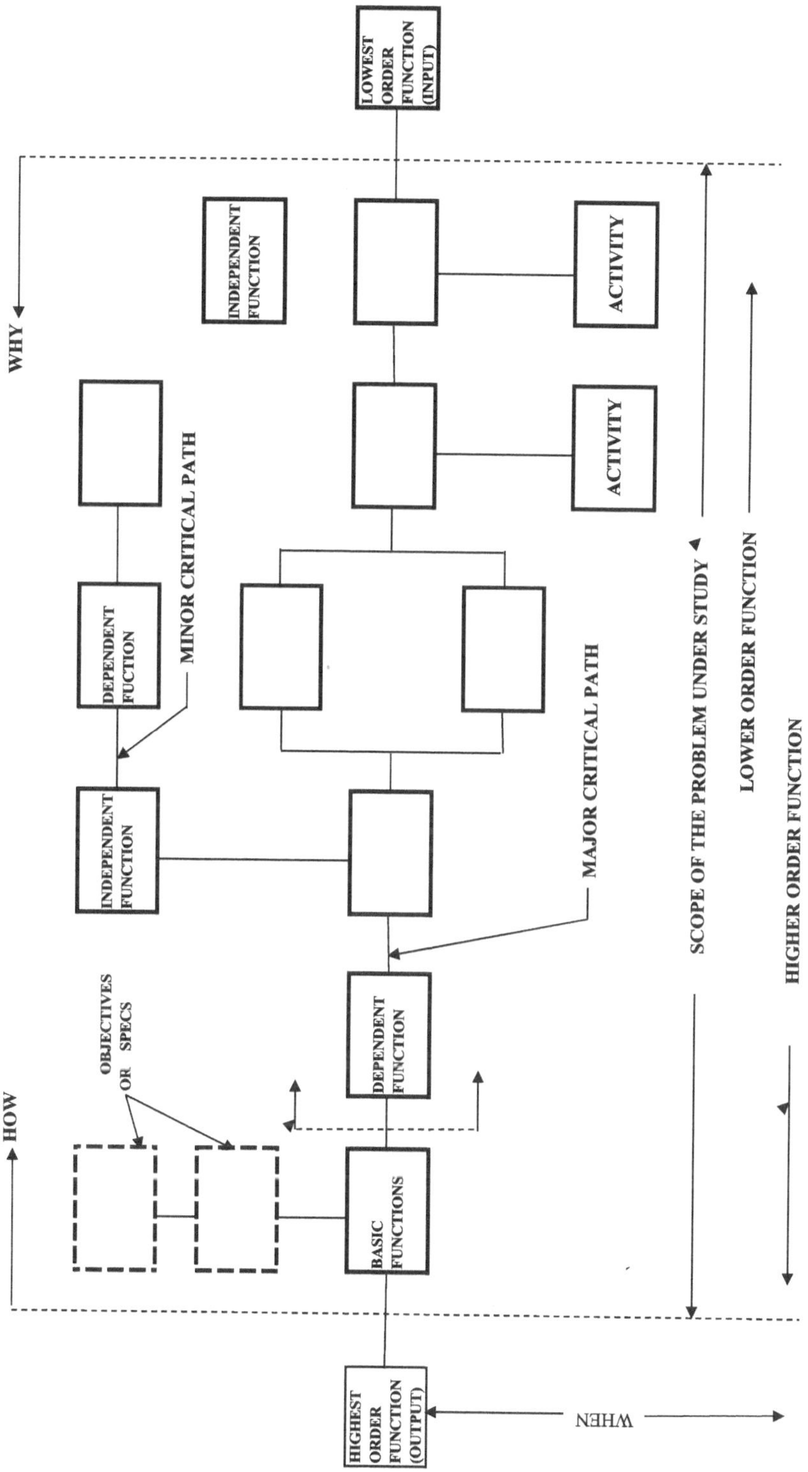

FIGURE 9.4 FAST model.

Bibliography

American Society of Civil Engineers (2000), Quality in the Constructed Project: A Guide for Owners, Designers, and Constructors, Second Edition, American Society of Civil Engineers, Reston, VA.

Alphonse J. Dell'Isola (1997), Value Engineering: Practical Applications for Design, Construction, Maintenance and Operations, RSMeans, Kingston, MA.

The Engineers Joint Contract Documents Committee (EJCDC) (2007), from the website http://www.ejcdc.org

Federation of Internationale Des Ingenieurs-Conseils (FIDIC) from FIFIC website http://www.fidic.org

International Organisation for Standardisation, from ISO website http://www.iso.org

Master Format (2014), The Construction Specifications Institute and Construction Specification, Canada, from the website http://www.csi.com

New Engineering Contract (NEC) from the website http://www.nec.org

Abdul Razzak Rumane (2013), Quality Tools for Managing Construction Projects, CRC Press (A Taylor & Francis Group Company)

Abdul Razzak Rumane (2017), Quality Management in Construction Projects, Second Edition, CRC Press (A Taylor & Francis Group Company)

Abdul Razzak Rumane (2019), Quality Auditing in Construction Projects, Routledge (A Taylor & Francis Group Company)

Abdul Razzak Rumane (2021), Quality Management in Oil and Gas Projects, CRC Press (A Taylor & Francis Group Company)

Save International Website: value-eng.org

Index

D

E

P

Q

R

U

V

W

Z